Spooky Action at a Distance

Felicia Watson

Published by:
D. X. Varos, Ltd
7665 E. Eastman Ave. #B101
Denver, CO 80231

Book cover design and layout by, Ellie Bockert Augsburger of Creative Digital Studios.
www.CreativeDigitalStudios.com

Cover design features:
3D illustration tunnel or wormhole, tunnel that can connect one universe with another. Abstract speed tunnel warp in space, wormhole or black hole, scene of overcoming the temporary space in cosmos. By rost9 / Adobe Stock
Spaceship on Black with Blue Engine Glow - science fiction illustration By Algol / Adobe Stock

ISBN: 978-1-941072-64-6

Printed in the United States of America

For Ed and Dave — thanks for the tremendous support, I couldn't have done it without you guys!

Table of Contents

Chapter 1

The Space Between

"There is fiction in the space between
The lines on your page of memories…
There is fiction in the space between
You and me" – Tracy Chapman, Telling Stories

"I hope you know what you're doing," Con said, as the view screen showed them speeding towards a thicket of jagged crimson rock spires in Centauria's Kādivi Valley.

"There's a first time for everything," Deck answered while adeptly weaving in and out of the obstacles at a blistering pace.

"Is there a good reason you need to be going *this fast*?"

"The best reason of all – I'm trying to break the record for this course."

With a skeptical glance at the power indicator, Con admonished, "Really hard to beat anything when you run out of fuel. And you still have the Devil's Eye to get through."

"You worry too much; those indicators always leave a little in reserve." She tapped the transdermal ship link on her forehead. “Besides, she’s telling me she *wants* to make it.”

“Uh...the ship actually talks to you through that thing?”

“Oh yeah, we are as one.”

Kennedy looked at the Blue-Heeler lounging in-between them on the floor. “Kayatennae, you may have a rival for best-friend status.”

“Don’t be silly; nothing could replace Kay. Besides he’s not my dog; he’s a—”

“—Corpsman, third class, yes, I know.” Con gripped the panel, warning, “Watch out for that—”

“I see it,” Deck said, deftly steering the ship under the stone outcropping. “I knew about it long before you did.”

“Because the ship told you?”

“Yes.”

“I’m not sure I like the idea of putting my life in the hands of a Nav-unit.”

Sailing through a canyon just above a raging orange river, with barely a meter to spare on either side, Deck countered, “It doesn’t work that way. Your life is still in *my* hands.”

“Oh, much better!”

Minutes later Decker was skimming through the long, narrow tunnel, known colloquially as the Devil’s Eye. With both daylight and the course record in sight, she assiduously ignored the low fuel warning the ship was giving her. Suddenly all the lights on the control panel went out and the cockpit went dark. She exclaimed, “Son of a *bitch*!” and then looked at Con, lamenting, “Not good.”

“Does this mean that—”

“Yeah, we’re fucked.”

A booming voice echoed in the cockpit. "Who's in here?"

Deck relaxed slightly upon hearing the question. "Maybe not so fucked after all." She stripped off the transdermal link and hopped out of the pilot's seat, smiling broadly at the imposing figure of Commander Pika Kelekolio. "Sir, good to see you. It's been much too long." Her smile faded as she recognized Lt. Commander Talako Jacoway at his elbow.

Deck nodded at him. "Commander Jacoway."

Jacoway sighed and snapped, "I might have known it was *you*."

In the meantime, Kelekolio had advanced into the Astronautics Lab and engulfed Decker's proffered hand in both of his huge ones. "Naiche Decker! You haven't changed a bit." He chuckled as he thumped her on the back, saying, "Not in any way. If you wanted some time on the simulators, Deck, all you had to do was ask."

"Not on the QNS-beta simulator, Commander," Jacoway said. "I would *never* have given such permission. Which is why I'm assuming Lieutenant Decker has been breaking in—"

"I didn't break in!" She grinned at Kelekolio, explaining, "You haven't changed the code since I was a cadet, sir. It's still your grandmother's birthday."

"I'll be damned," Pika proclaimed, his tattooed face alight with glee. "You remembered that?" He said to Jacoway, "I'm sure Deck did your simulator no harm, Tal. No need to make a global case out of this, is there?"

"It needs to be reported to her commanding officer, at the very least."

"Consider it done," Con announced, as he emerged from the simulator, accompanied by Kay. He nodded at the group. "Lieutenant Commander Conroy Kennedy, at your service."

"I see," Jacoway said. "Then perhaps you'd be so good as to give me the name of *your—*"

"Commander Nils Lindstrom. His IM designation is—"

"I'll look him up."

Kelekolio stepped in between them saying, "I don't believe we've met before, Kennedy."

"No, sir, we haven't," Con replied. "I opted out of Flight as a cadet."

"You did all right without it though. I've heard plenty about your exploits on the *Lovelace*." Kelekolio leaned down to give Kayatennae a pat on the head. "And his...." He looked at Decker confirming, "This is *Lovelace's* famous Search and Rescue dog?"

"He is indeed."

Kennedy said, "I finally get to thank you for gracing my Tactical-Front squad with one of the best Micro-craft pilots ever."

"You were Deck's CO at the front, too?" Kelekolio winked at Decker, while saying to Con, "I'm not sure if I'm more envious of your luck or your endurance."

His dark brown eyes alight with mirth, Kennedy answered, "I think a combination of the two is the most appropriate reaction, Commander."

"Well, I wish I had time to buy you a beer and hear the stories, but I've gotta be going. Got some student exercises to program." He turned to Jacoway who had been visibly fuming during that genial exchange. "It's up to you, Tal, if you wanta report a coupla war heroes for a bit of harmless fun." He strode away, directing, "Lock up before you leave."

"I will, sir. Though I strongly suggest changing the entrance code!" Jacoway called after him. He faced Naiche and Con in silence for a moment before scoffing at her, "Still using sex and your surname to get by in the

UDC, huh, Decker? And I hear *daddy's* now running interference for you, too."

"That was simply Commander Kelekolio displaying his usual comradery and good humor." Naiche tried to resist tacking on a personal observation, but she didn't try hard enough to remain silent. "Though seeing how completely lacking you are in both of those, I'm not surprised you didn't recognize them."

"Are you forgetting I'm your superior officer, *Lieutenant*?"

"No, sir. That's the only reason you're still standing after that 'sex and surname' crack."

Jacoway appeared indifferent to the implied threat, but his response dripped with disdain. "Is physical violence your answer for everything?"

"It has always served me well. Especially at the front. But I know prima-donna fighter pilots never had to learn anything about *that*."

Tal's hands balled up at his sides and he seemed to be struggling for a reply. At last, he crossed his arms and observed coldly, "Your late mother would be so proud of your self-proclaimed bellicosity."

Naiche felt Con stiffen beside her and surreptitiously nudged his shoulder with hers to let him know it was okay. She knew Jacoway was aching for her to lose it, for an excuse to report this infraction without looking small-minded. She instinctively gripped the moonstone locket around her neck with her right hand. One deep breath, then another enabled her to reply with relative calm, "Yeah, maybe I'm not a diplomat like my mother, but she'd still take comfort in knowing I never resorted to *a low blow*."

Jacoway opened his mouth and then closed it, finally drawing himself up to his full height – which was barely an inch taller than Decker – to say, "I believe you know the way out. Now would be a great time to use it."

When they were clear of the Astronautics Center, Con asked, "What Jacoway was insinuating back there about you and Kelekolio.... You two were...involved?"

"Yeah, in my fourth year. *Well after* I was done with Flight basics," she assured him. "A brief but memorable liaison."

"Uh-huh. And what's the deal with Jacoway?"

"He's just some asshole I had a couple of Astro-nav classes with. We didn't...umm...we butted heads. A lot."

"Oh no, he's not another secret relative of yours, is he?"

"No! Yuck, why would you even say that? You thought he was Chiricahua?"

"*No.* Now I just automatically assume you're related to anyone you have a long-standing and mysterious beef with."

"You met the man – do you really think me having a problem with him is *mysterious*?"

"I still wanta hear the whole story." When Deck didn't respond, Con insisted, "Now."

"Okay," she sighed, "the *whole* story's kind of long and convoluted, but I can give you the bottom line upfront." Naiche headed over to a low stone wall surrounding one of the many gardens dotting the Uniterrae Defense Corps Academy campus. She parked herself on it, cross-legged. Con followed and sat next to her, stretching out his long legs. Deck glanced around before explaining, "Jacoway's people are part of the One Nation Consortium—"

"Isn't it the One Nation Collective?"

"Yeah, whatever." Naiche drew a deep breath and then continued, "He claimed my band of Chiricahua are isolationists—"

"Well, in a manner of speaking—"

"We are *independent*! There's a difference."

"That's it? That's the whole problem between you two?"

"Isolationist was one of the *least* patronizing things he said about my people. He claimed that our ways are archaic...and that we're dying out. So, one day I showed up at an off-campus talk he was giving to some local *Ndee* about the Consort– Collective, talking about how First Peoples are better off as one, extolling everything that his band of Choctaw had gotten from being part of it."

"And?"

"And I asked him a question. 'Do you speak Choctaw?'"

"Doesn't seem so bad."

"Well...." Deck drawled, with a slight smirk. "It was probably the way I asked it. '*Chahta imanumpa ish anumpola hinla ho*?'"

"Let me guess – that's Choctaw?" When she nodded Con asked, "Did he know what you asked?"

"Nope." Naiche snickered as she explained, "He said he didn't speak Chiricahua. I had to enlighten him to the fact that he couldn't even *recognize* his own language – let alone speak it."

"Deck, you humiliated him – in public."

"No, I made a point! That the Collective is a farce – if you're *everything*: Navajo, Cherokee, Choctaw, Osage, Paiute – then, you're *nothing*!" She added, more calmly, "Humiliating him was merely a bonus."

"Remind me never to get on your bad side."

"When it comes to *you*, I don't have a bad side."

Kennedy smiled at her assurance and ventured, "Let me take another wild stab here. That one course record you're so determined to beat? Does Jacoway hold it?"

"Yep. No one else has even come close. And every freakin' time I try—"

"You flame out?"

"Yep. But I'm getting closer. I just need some more time on that damn simulator."

"Tell me you're not gonna try again. No, let me make that an order. You are *not* going to try again."

"Don't worry – I couldn't even if I wanted to. Knowing I was at it, he'll lock it up tighter than his own ass."

"Will he guess what you were up to?"

"He will if he looks at the data."

"And how ticked off will that make him? Enough to rat us out to Lindstrom?"

"I don't know." Naiche rocked her head slowly from side to side as she considered the question. "Could go either way. I'm sure he doesn't wanta look petty, but he'd love making trouble for me."

"Wait, was he one of the fighter pilots you and Ato were jawing with that one time? In that bar on Neruda?"

"Yeah, he was one of them."

"Didn't it almost come to blows?"

"Almost...but then they backed down." With a short laugh, Deck elaborated, "Peacocks not being big on hand-to-hand combat."

"Just great," Con groaned.

"Oh, what're you worried about? If he turns us in, you'll get three minutes of tsk-tsking from Lindstrom while I'll get three days of scolding from Ricci." Her tone switched to a sing-song mimicry of her father. "'We've talked about this, Naiche. I'm disappointed in you, Naiche. I expect better from you, Naiche.'" She ran her hand through her long brown hair and mused, "Maybe it would be better if I just told him myself."

Kennedy cocked his head, stating, "But Jacoway might not say anything. You think it's worth the risk?"

"If I can catch Ricci in a really good mood, it might be."

"You ever replace his grappa?"

Decker winced slightly. "Not yet." When Con shook his head in exasperation, she explained, "That stuff is expensive!"

"No shit! Probably why he was so pissed."

After a moment's reflection, Deck exclaimed, "Hey! He's seeing Stein tonight. I'll tell him tomorrow morning."

"I'm not following you. Why tomorrow morning?"

"*Because* he's guaranteed to be in a good mood. Sex puts *everyone* in a good mood."

"Yeah, but what if they're just having dinner or something?"

"You're hilarious."

"What?"

"He doesn't hang out with Admiral Ball-buster for her pleasant company."

Con laughed, retorting, "For your sake, I hope you're right about this."

"Yeah, me too."

***

Early the next morning and fresh from their run, Decker and Kay bounded into the quarters she shared with her father, Captain Matteo Ricci. She saw Ricci standing at the galley kitchen's counter, pouring a cup of coffee.

"Hey, Pop, I'm glad you're up. There's something I wanted to talk...to...you...." Naiche trailed off as she realized that Ricci was wearing the same clothing that he'd left in the night before. "You're just getting in?"

"Yes." Ricci slid stiffly onto one of the navy-leather stools and hunched over his coffee. "And good morning to you, too."

"Yeah, sorry, good morning." She entered the kitchen and started scooping dog food into Kay's bowl. As she placed the bowl in front of the dog she confirmed, "You spent the whole night with Stein?" There was no way to disguise the surprise in her query.

He looked slightly annoyed at the question. "Yes, I spent the night at *Admiral* Stein's place. Is that a problem?"

"No, it's just kind of...*unusual.* Isn't it?"

"I suppose it is."

As she helped herself to some coffee, Naiche ventured, "Bet there's a story there."

"In a manner of speaking."

"Are you gonna share it?"

"No, I'm not."

"Because you're my father or because you're my captain?"

"Yes."

Naiche rolled her eyes and got a container of leftover chili out of the fridge. She could feel Ricci watching her as she stuck it in the auto-cooker and got a box of crackers out of the cabinet.

"Is that what you're having for breakfast?"

"Yeah, you want some?"

Grimacing at the suggestion, Matt declared, "*No,* thanks."

"Well, there's not much else to eat. We didn't get a food delivery this week."

Ricci exhaled and swiped a hand across his forehead. "Oh, I guess our renewals ran out." He raised his eyebrows at Naiche.

"Hey, don't look at me. I'm just a boarder here. That's not my name on the on the front door."

"Well, Boarder, how about earning your keep by placing a request for immediate delivery?"

"Sure thing." Decker pulled out her hand-held computer and started entering the order.

"Get three days' worth."

With deceptive nonchalance, she ventured, "Three days, huh? Guess you're expecting us to ship out soon?"

"Pretty soon."

"Which means this mysterious mission of ours—"

"Stop fishing. You'll hear all about our mission when it's fully approved. Then, and only then."

Naiche heaved a dramatic sigh, but, recognizing the futility of further digging, she changed the subject. "You want me to have Mess send over a hot breakfast for you while I'm at it?"

"That'd be great. Get me my usual. A spinach—"

"—and provolone frittata with fried potatoes. I know."

Ricci downed his remaining coffee and asked, "Now, what'd you want to talk to me about?"

"What?"

"When you came in, you said there was something you wanted to talk to me about."

"Oh yeah...*that*. It's nothing important. It can wait until you've...." *Had a mood enhancer,* she privately supplied. "Eaten."

"Okay, then, I'm gonna grab a shower. Have Mess deliver that frittata in twenty."

"Will do."

Before Ricci made it down the short hallway to the bathroom, Decker heard his hand-held buzz and held her breath, hoping it wasn't a message from Lindstrom. When he strode back into the living room and addressed the AI unit on the wall, "VICI, arrange a meeting with Commander Lindstrom in my office, 1000 hours," her heart sank. She braced for the onslaught of a blistering lecture but Ricci continued, "And then set up a *Lovelace* senior staff meeting, in the adjacent conference room, for 1030 hours."

"Geeze, that's kinda over-kill, isn't it?"

"What's that mean?" He stared at her a second before clarifying, "I just got a message from Fleet Admiral Bindroo. Our mission orders have been approved."

"Oh!" Deck quickly tried to hide her extreme relief. "Awesome. Now, can I have the details?"

With a sage smile and a shake of his head, Ricci turned back to the VICI unit. "Ensure that Commander Charani is available to meet with us at 1030."

"What does Zache have to do with our mission?"

"You'll be fully briefed at 1030 hours, with the rest of the senior staff."

"You're no fun," she groused as Ricci headed back towards his shower.

He tossed back, "That's definitely NOT what she said."

"Oh – *that* you can tell me!"

Chapter 2

Every Man Be Blind

*"The Truth must dazzle gradually
Or every man be blind" – Emily Dickenson, Tell All the Truth but Tell It Slant*

Ricci looked at his senior staff gathered around the conference table with a smile and announced, "I'm not sure introductions are strictly necessary. If I'm not mistaken, you all know Commander Charani. Most from either being a cadet with her or taking Theory of Astronavigation from her." There was a murmur of agreement from the group. "Zache, I hope no one here has reason to squirm. Did they all pass?"

"No worries on that score. In fact, you have two of my best pupils ever," she replied, gesturing towards *Lovelace's* pilot, Tanja Petrović, and then to Decker who, as always, was seated beside Kennedy. Next to him was his wife, Lt. Commander Aqila Lateef, the ship's chief science officer.

“I can certainly return the compliment, Commander,” Lieutenant Petrović said. “You were my favorite instructor. I’m sorry to hear you’re not teaching anymore.”

“Oh, well, no matter how much you love something, everyone can use a change after twenty-some years.” She turned to Ricci and said with a laugh, “Don’t tell Nik I said that.”

Matt wondered if Zache was referring to the fact that she and Lt. Commander Nikolaos Sinos had been married for twenty-seven years or the fact that he was still teaching history and diplomatic studies at The Rock and would inevitably continue doing so until they hauled him out feet first. He brushed the thought aside and turned to the business at hand. “Commander Charani will be accompanying us on our mission to the Okeke Phenomenon.”

“Excuse me, Captain,” said Jeff Sasaki, the ship’s linguist. “I’m not conversant with all of the many interstellar phenomena. Which one is that?”

“That’s the quantum entanglement identified by Doctor Okeke last year.” Seeing that Lt. Commander Sasaki still looked slightly doubtful, Ricci clarified, “Commonly known around here as the Expanse.”

“AKA the Death Trap,” Decker supplied.

Ricci shot her his most quelling glare. “Not. Helpful. Lieutenant.”

She seemed suitably admonished and offered a contrite, “Sorry, Captain.”

Lateef asked, “Is this another rescue attempt?”

“Yes, it is.” Matt pressed a button on his control panel and brought up a holographic screen showing the details of the Expanse and a schematic of a ship. “The *Burnell*, a scientific vessel, went out to explore the Expanse over two months ago. They quickly discovered that due to the ever-changing quantum state of space-time in that area, navigation proved...difficult."

Chief Engineer Carla Ramsey interjected, "Yeah, how can you navigate a space-time field that's in constant flux?"

"Turns out you can't," Charani answered. "Not with our standard Nav-sat."

Matt changed the display to the next screen, showing another ship and expounded, "Fortunately, they were able to get a distress call out to a relay station in the short time that they were near the margins of the Expanse. When that transmission reached the UDC, the Fleet Admirals immediately sent out a rescue vessel. That was the *Meitner*, which has not been heard from since it entered the Expanse and is now presumed to be similarly...lost there."

"Refresh my memory," First Officer Nils Lindstrom, said. "Ours will be rescue attempt *number...?*"

"Three," Ricci replied. "The most recent attempt was by a ship, the *Daleko*, equipped with an experimental quantum-computing assisted navigation drive, designed by Commander Charani."

"Quantum assisted computer...you want to run that by us again?" asked Dr. Rita Clemente, the chief medical officer.

"We just call it the Quantum-Navigation System or QNS," Charani explained. "The QNS can perform seven-thousand exaFLOPS, constantly re-adjusting the ship's direction to keep it on course. Basically, the QNS can keep up with the instability of the surrounding quantum entanglement."

Lateef asked, "If that's true, then what happened to the *Daleko*?"

Ricci answered her. "Its fate remains at this point...unclear. Communications from the *Daleko* have been retrieved and they indicate that the ship made it into the Expanse and that the QNS drive worked well but shortly afterwards they were forced to abort. Those later communications were so garbled, that we're not sure *why*. An early interpretation of their last message was that they might have been under attack."

As chief tactical officer, that information commanded Kennedy's immediate attention. "Attack! By what? Or should I say by who?"

"We're not sure. Of that or anything else, really. There has been no further communication from the *Daleko*. We don't know if they were actually attacked, what shape they're in, or where they ended up." Ricci paused to punctuate the gravity of his next statement. "There's even some concern that whatever attacked them might have...followed them out."

Decker said, "Oh. *That's* why they're sending a command ship this time."

"Exactly."

Petrović learned forward, eyes alight. "Is *Lovelace* going to be equipped with a QNS?"

Smiling at her obvious enthusiasm, Ricci shook his head. "Sorry to disappoint you, Petrović, but no, *Lovelace* will not get a QNS. We'll be ferrying the vessel, the *Cerxai,* that has the QNS – and we are under strict orders to remain *outside* the Expanse." Ricci noted that the group showed mostly relief at that pronouncement.

Ramsey asked, "Is the QNS system the one that requires a neural link between the pilot and the Nav-unit?"

"That's right," Charani replied. "It's the only way for a human pilot to keep up with the QNS."

"How does that work? I heard that the pilot has to get a broadcast port inserted in their head. Can that be right?"

Decker broke in to assure Ramsey, "Nah. It's just a transdermal band that you...umm...wear." Ricci stared at her with narrowed eyes wondering how she'd acquired that particular bit of knowledge. Detailed info about the QNS system had been restricted to a need-to-know basis. Most heads had turned towards Naiche, waiting for her to continue; she simply mumbled, "I mean, that's what I've heard. You know...pilots gossip...after a few drinks."

"No, Decker," countered Zache, "Commander Ramsey is right. The transdermal band is for training exercises only. Going live on the production QNS *does* require that a broadcast neural port be inserted."

Decker exclaimed, "That's why I couldn't—!" She stopped abruptly, not finishing her thought and instead, looking at Kennedy.

Matt knew Naiche was hiding something but wasn't sure he actually wanted to know what it was. Still he felt compelled to ask, "You couldn't what, Decker?"

"Umm, I couldn't really understand what those pilots were talking about...you know...at the bar."

Lindstrom observed acerbically, "Just to be clear, when you say, 'neural port,' you are talking about drilling a hole into someone's skull? In order to insert a computer interface?"

"It's quite small and located behind the ear," Charani answered. "You really wouldn't notice it, unless you knew what you were looking for."

"Yes," Nils drawled. "Because that would be my greatest concern if the UDC was ramming a computer port into my brain. Whether or not it ruined my aesthetic."

Petrović asked, "Are you going to be the QNS pilot on this mission, Commander?"

"No, my specialties are theory and design; I'll be riding shotgun on the *Cerxai*. The pilot selected for this mission is someone who's been helping me to refine the QNS over the past few weeks. I think some of you know him – Lieutenant Commander Talako Jacoway."

Matt saw the look of concern pass between Decker and Kennedy and assumed that his Tactical team wanted more information about the attack on the *Daleko*, but he had none to give.

***

After the mission briefing, Ricci and Charani were heading to the mess hall to have lunch together when Matt stopped to stretch his aching back again.

Zache took the opportunity to tie back her mass of curly dark hair, which had been blowing into her face, but asked, "Is something bothering you? That's the third time you've stopped. I've seen you sprint across campus without this much trouble."

With a grimace, Ricci explained, "Ah, I wrenched my goddamn back last night. Could barely walk until early this morning. I'm still feeling it."

"Medical couldn't help?"

"I didn't deem it necessary to involve Medical."

"Oh," Zache said. An impish smile lit her dark complexion as she asked, "Did this by any chance happen during sex? That up-against-the-wall maneuver you told me and Nik about?"

"What makes you say that?"

"Am I wrong?"

After a moment's hesitation, Matt was forced to admit, "No. How did you guess?"

"We're getting to that age, Matty."

"What's this 'we' shit?" Ricci chuckled. "You're a year older than me."

"So, I would know, Old Man." Matt thought the use of that sobriquet had earned Zache a full-bore scowl and he directed it her way. "Hey," she retorted, hands aloft in mock innocence, "that's what they call the captain of a ship, isn't it? 'The Old Man'."

"Not to their face."

"Don't I get to take a few liberties?"

"A few – but best not to call me Matty on the bridge."

"This is gonna be weird, isn't it? Me under your command for the duration of this mission."

"Hey, weird is in my wheelhouse – my daughter's part of my crew."

"I know – and it must be hard having to dress her down in public."

Matt shrugged, replying, "I haven't had to do that in a while.

"What about this morning?"

"What? That death trap thing?" He looked at Zache in surprise. "You consider that a dressing down?" Matt scoffed. "I gave her worse than that when her boyfriend drank the last of my hundred-year-old grappa."

"Ah, yes," Zache, observed archly. "She's under your thumb at work – *and* at home."

"Under my thumb! What the hell does that mean?"

"Nothing. It was just a joke."

"I guess I'm not seeing the humor there."

Rather than explaining the intent of that zinger, Zache zagged back to their recent topic. "Is that why she broke up with Diego? Because he drank your grappa?"

"No," Matt answered, after taking a moment to adjust to the abrupt change of topic. "She said that relationship had simply run its course."

Charani, nodded and then ventured, "Speaking of men Naiche has problems with...."

"Yes?"

"Did you notice how she reacted to the mention of Tal Jacoway?"

"I noticed she was...hiding *something* this morning – but nothing about Jacoway. Why?"

"They don't get along."

"How do you know?"

"They were in my Astro-nav classes together."

"They were? I could have sworn Jacoway was a little older than Naiche."

"He is. She sailed through all of the piloting labs so fast, we had to keep advancing her to maintain her interest. Like you, she ended up taking Astro-nav with the upper-classmen."

"She did?" Matt confirmed with a smile – and a bit of a swell to his chest. "Gordon told me she struggled as a cadet."

"She did struggle. In almost everything but piloting and flight classes. At the time, we all found that surprising but now, of course, it makes perfect sense." Ricci felt the familiar weight of guilt when it came to the subject of his daughter's long-hidden paternity. Twenty-nine years of friendship must have enabled Zache to read his moods because she went on to gently add, "You know, she suffered a bit from language issues but mainly from unfair expectations. Maybe it was best that she didn't have the dual burden of being both a Decker *and* a Ricci back then."

Matt appreciated Zache's assurance but it was too raw a subject for him to want to continue, so he asked, "What was the issue between Naiche and Jacoway?"

"I never really figured that out. I first noticed something when I assigned them as study partners and a few days later Naiche asked to be reassigned. And then when I paired them for a senior project, she and Tal both objected, claiming they didn't 'mesh'. The only group open to Naiche by then had my two worst students. She ended up doing all the work herself and still managed to come in second to Tal's group." Zache laughed and said, "That girl is made of iron."

"As was her mother."

"Is that you being modest?"

"No, that's me being honest."

"Then be honest and tell me why you accepted this mission as is, when last week you had doubts about proceeding with...." Zache paused and faced him head on. "...the B-Team."

"Damn it, Zache! I never called you and Jacoway the B-Team. I know he's a gifted pilot but Jackie Bastié is the most decorated pilot the UDC has ever produced. And while you designed the QNS-drive, Stefan Pilecki built the fucking thing. And we still have *no* idea why they decided to abort the mission so quickly."

"They were attacked."

"Not if the latest sequencing of communications is correct. The attack, if there even was one, came during the retreat. I just wanted to know more about what went wrong before we rush headlong into yet another disaster."

"What changed your mind?"

Since they had reached the entrance to the mess hall, Matt hesitated. He took a deep breath and exhaled slowly. With a jerk of his head he indicated that Zache should follow him over to a secluded alcove. After confirming there was no one within earshot, Matt said, "This is not general knowledge at this time. I don't know if it will be before we ship out." Zache looked concerned but simply nodded. "The astrophysicists in Scientific think the Expanse is collapsing."

"Collapsing?"

"Rearranging, refolding...I don't know what you call it but they think it's gonna disappear from our sector of the galaxy."

"And those eighty people trapped inside?"

"I'm told that they'll probably end up hundreds of thousands of light years away." Matt let Zache digest that news briefly before saying, "We're their last hope, Zache. And we've got to move fast."

***

The next afternoon Ricci showed up at Charani's office for a meeting with her, Jacoway, Ramsey, and Lateef. He was a bit early but after a brief knock and hearing no objection from Zache, he walked in and found her having a holo-chat with her son, Teo.

The young man recognized him immediately and offered a warm greeting. "Hey, Uncle Matt. How's the space biz?"

"The space biz is busy. How's Rigelkent treating you?"

"It's different for sure. But I'm getting used it. Meli's helping me settle in."

"Ok, hon," Zache said. "I have a meeting now so I'll have to sign off. I'll talk to you when I get back Earth-side."

"Oh, yeah, Uncle Matt, I hear my mom is running off with you."

"Yes, my life-long dream has finally come true," Matt laughed.

Teo snickered in return and gave Ricci a thumbs-up. He turned back to his mother, calling out, "Ok, bye! Love you, Mom!"

As the hologram faded from sight, Zache swiveled her chair to face Matt. "Your life-long dream? You liar."

Matt shrugged. "I don't know, do you ever wonder what would have happened if we'd met first? Before you met Nik and I met Naomi?"

"I think we'd still be nothing more than good friends." While Matt was nodding in agreement, Zache added, "Which is for the best because any kid of ours would've had a nose so big they wouldn't've been able to walk upright."

"Sounds about right," Matt admitted with a grin. He nodded towards where Teo's image had been, observing, "It seems like the twins are doing well."

"Yeah, they're doing great. The Centauri settlements are the place to be these days. Since you ended the war, that is."

"Oh, stop with that," Matt protested as he waved a hand at her and settled into one of the ancient chairs at her small conference table. "Why are you still in an Instructor's office? Weren't you suppose to move to the Engineering Center?"

"That's what I've been asking. Marijke Bengtsson keeps telling me they're *working on it.*"

"Yeah? Well, next time she stonewalls you, tell Beng you didn't get that fourth stripe for its decorative purposes."

"Who the hell is reckless enough to talk to her like that?"

"I am. On occasion."

"Huh, I guess a Founder's medal winner can get away with it. I talk to Beng like that and I'll end up with an office in the cadets' bathroom."

Ricci was only half-listening since he was pondering whether or not to broach Zache's remark of the day before. It had nagged at him all night, so he decided he had to. "Hey, can I ask you something?"

"You can ask me anything."

"What did you mean yesterday about Naiche being 'under my thumb' at home? And don't tell me that it's just a joke again – that joke came from somewhere."

"Okay, fair enough." She ran a hand over her chin and said, "I *have* been wondering if I should say something...."

"If you can tell me I'm getting old, you can tell me whatever this is."

"Naiche's been a first lieutenant for a while now – why hasn't she gotten her own quarters?"

"Because she likes living with *me.*"

"She said that?"

"Not in so many words...." Doubt crept into Ricci's mind as he recalled Naiche referring to herself as a mere boarder in his quarters.

"Maybe she does. But I just wonder if a little space wouldn't do you *both* some good. While I know you're trying to make up for lost time, Matt – you can't do that by trying to make her a little girl again."

"You think *that's* what I've been doing?"

"Not consciously, no – but subconsciously? Maybe."

Before Ricci could respond there was a knock at the door announcing the arrival of Jacoway. After Zache introduced them, Jacoway shook Matt's hand and said, "It's an honor, Captain."

"I know I've seen you around, Jacoway." Ricci, answered. "You were with Commander Coleman's Squadron during the war, right?"

"Yes, sir."

"She's a hell of a pilot, isn't she?" When Jacoway nodded eagerly, Matt added, "Jess and I flew together in Astra-Six."

Tal's eyes widened and his mouth opened slightly. After a second, he confirmed, "You were a fighter pilot, Captain?"

"Hard to believe, huh?"

"No, it's just that...does your...I mean...do most people know that?"

"I never thought about it. Maybe not; it was only for a couple of years and then I went into Command Operations."

"They couldn't wait to boot Matt up the ladder," Zache said.

"Sorry I'm late." The group turned to find Aqila Lateef standing at the open door. She walked in explaining, "I was just messaging with Commander Ramsey. She's not going to be able to make the meeting."

"Something wrong?" asked Charani.

"No, she's aboard the *Lovelace* and can't get away. You know what preparing for launch is like."

Zache smiled wryly and admitted, "Not really."

"Oh. Well, neither did I a few years ago." Aqila sat down and pulled out her hand-held. "I promised to take detailed notes for Carla. She's extremely interested in how this virtual tether will work."

"Okay, then, let's get started." Zache walked over to the rather archaic holo-screen on the wall and brought up a schematic. "We knew that once the *Daleko* located the lost ships, we would still need a way to guide the *Burnell* and *Meitner* out of the Expanse. The problem being that there was no possibility of either ship staying with the *Daleko* for any length of time. Our solution was to deploy a space buoy that emitted a gamma-ray resonance to the *Daleko,* which reflected it back. The resonance would be relayed back and forth between the two, forming a sort of 'homing beacon' that the other ships could follow along out of the Expanse."

"But they wouldn't be able to trust anything their Nav-sat is telling them along the way, right?"

"That's right."

"I take it that the risk of collision is considered low enough that it's not a concern?"

With a sigh, Zache answered, "I wouldn't say it's not a concern, just that it's pretty much their only way out, so, yes, the risk is worth it."

"Do we know if the tether worked with the *Daleko?*"

"Yes. The buoy and all the communications on it have has been retrieved. We can definitively say that the tether worked exactly as expected – until the *Daleko* broke the transmission during their retreat."

"Then why are we using the *Lovelace* to emit the pulse rather than a buoy, this time?"

Charani looked at Ricci. "Am I at liberty to discuss this?"

"If you're referring to the increasing instability of the Expanse," Lateef interjected. "Ramsey and I were briefed this morning."

"Good. The answer is that *Lovelace* can emit a stronger, steadier beam than the buoy, and we might be needing that."

Matt said, "That does mean you two will have to wait to exit the Expanse until the other ships have made it out. That could take a while." Matt looked at Jacoway. "And I understand that the QNS puts quite a strain on the pilot."

Jacoway answered, "That doesn't worry me, sir. I've trained extensively...and I have complete faith in the QNS."

"And I have complete faith in you," Charani said.

"With all this good faith, how can we fail?" Lateef exclaimed brightly. She stood up saying, "Send all the necessary specifications to my in-box, Commander, and my team will work with Engineering and have the tether ready to go within a week."

Matt remained uncharacteristically silent.

Chapter 3

Often Born of Blood

"You must remember, family is often born of blood, but it doesn't depend on blood." Trenton Lee Stewart, The Mysterious Benedict Society

The day after mission launch, Commander Lindstrom was escorting Charani and Jacoway to Captain Ricci's office aboard the *Lovelace* for an early morning breakfast meeting. The group was walking down the passageway when they heard Kennedy's voice call out, "Gangway!" and a running trio, consisting of two humans and a dog, flashed by. Tal recognized the group as the same one he'd caught making unauthorized use of the QNS simulator earlier that week.

Lindstrom said, "Just as an FYI, that...err...activity takes place throughout the ship's passageways every morning about this time."

Charani asked, "Doesn't *Lovelace's* gym have virtual tracks?"

"Actually, yes, we do. When they first came aboard, I did ask why they weren't making use of them. But Kennedy and Decker explained their aversion so *eloquently* that the captain was moved to allow it." The first officer put a hand to his face, saying, "Let me see if I can remember the exact phraseology. Ah, yes." With a raised eyebrow, he looked down at them from his towering height and quoted, "'Virtual tracks are for pussies.'"

"It seems slightly...disruptive," Jacoway observed. "No one has ever complained?"

"A few people have." Lindstrom shrugged. "My ultimate decision was that when the complainers defend this ship and its crew as ferociously as do Kennedy and Decker, I'll put a stop to it. Until then...." Nils raised a languid fist. "Gangway!"

Tal wondered how much of that decision was predicated on the fact that one of the runners was the captain's daughter. Since boarding the *Lovelace*, he had observed that the two spoke to each other with the same degree of formality as would any captain and lieutenant but still found the situation odd and, considering his history with Decker, unsettling.

The subject was still on his mind when he entered shuttle bay-3 later that day to find Decker lurking about. He figured the wayward lieutenant was sniffing around the *Cerxai*, which was being housed there, and decided he wasn't going to let her get away with it. "Any particular reason you're hanging around this shuttle bay, Lieutenant?"

Turning at the sound of his voice, Decker, to his major irritation, smirked at him. "Just finishing up a weapons' check of the equipment lockers here, Commander."

"And why would the ship's pilot be doing that?"

Decker pursed her lips and stroked her chin while musing, "I don't suppose she would be...." Her voice strengthened as she declared, “But the senior Tactical officer might.”

Tal’s eyes narrowed in suspicion and doubt. “You’re still a Tactical officer?”

“I can see that you didn’t read the *Lovelace* briefing packet.” After he shrugged in response, a silent admission that he hadn’t thought it necessary, Decker asked, “Why wouldn’t I still be in Tactical?”

“I just assumed you would have moved to ship’s pilot after the war ended.”

“You ever fly a command ship?”

“Can’t say that I have.”

“It’s about as interesting as...well, playing that *N’daa* game, chess – if you’ve ever done that.”

“I have.” With a glare he added, “I happen to love it.”

Decker cocked her head slightly and with a small snort of amusement, said, “Well...of course you do.” She started walking towards the exit. “If you’ll excuse me, I have Tactical matters to attend to.”

As she passed him, Tal saw her covertly searching the area above his uniform collar. “It’s on the other side.” He turned to show her the small broadcast port inserted behind his right ear.

Decker advanced closer for a better look. “Is it very different from the transdermal link?”

“Night and day. You really don’t know the difference between the QNS and your own mind. Your hands just immediately know where to guide the ship next.”

“Wow.” Decker hesitated slightly and in a conciliatory tone offered, “You know...I never thanked you for not telling Lindstrom or Ricci about my little joy ride.”

“Joy rides.”

“Yeah, right, joy rides.” With a nod she said, “Well, good day, Commander,” and then turned to go.

“Did I miss the ‘thank you’?”

Decker spun on her heel, and pointed to where she'd been standing, stating, "Uh, that was it." She then resumed her progress towards the doorway.

"I'm over-whelmed by your gratitude."

"I'd volunteer to kiss your ass, sir," she retorted with a laugh, "but you turned that offer down years ago."

He watched her disappear from sight, cheating him of the possibility of a rejoinder. Not that he had one ready, anyway. As always, Decker left him fuming and speechless. He kicked a wrench laying on the floor in an attempt to relieve his frustration.

"Hey, what did that innocent wrench do to you?" Jacoway looked up to see that Commander Charani had entered the shuttle bay.

"Oh, it's just a poor substitute for Lieutenant Decker's behind."

"Are you two at it *already*?"

"Ah," he sighed, waving his hand, and explaining, "we have a history."

"I know." Zache came over and put her hand on Jacoway's shoulder. "Look, you may not think so, but if she's being insubordinate, Ricci would definitely want to know. And he'll put a stop to it. He doesn't let her get away with...well, anything, actually."

Tal shook his head. "No, it doesn't rise to the level of insubordination." He paused, trying to articulate his problem with Decker, finally stating, "She's purposefully difficult...and perpetually sarcastic."

"Oh, that's a bit of a tougher sell," Charani admitted. "If being difficult and sarcastic was actually considered to be a problem on this ship, I don't see how Lindstrom would ever get out of the brig."

"Yeah, what is *with* that guy? Has he always been like that?"

"Pretty much. I was in some Engineering classes with him at The Rock." With a laugh she recalled, "I remember once Commander Weingarten asking him if perhaps his mother's breasts had contained vinegar rather than milk." While Tal was chortling at the quip, Zache expounded, "But then again, Lindstrom's undoubtedly paid a price over the years for his caustic demeanor."

"How so?"

"It's held him back."

"He's a first officer!"

"Reporting to a captain two years his junior who *wasn't* a valedictorian." Having lost the thread slightly, Tal looked at Charani in confusion. "Like he was," she explained.

"What? Do you have all of The Rock's valedictorians memorized? I mainly remember the one who passed the medal on to me."

"Same here," Zache explained with a smile.

"Oh, I see."

Charani winked, and then pointed at the *Cerxai*. "Come on, let's give our baby a thorough once over."

***

The next day Decker was going over personnel weapons qualifications with Chief Corpsman Marvin Werther. As she dismissed him, she said, "And remember – I need all Enlisted assignments and shift designations in my in-box by 0900 tomorrow."

"0900?" he asked. "Can I get a little leeway on that?"

"No, sorry, you can't. I need to compile all of the personnel rankings so Commander Kennedy can file his Mission Readiness Status." She took a sip of her coffee before explaining, "Captain Ricci expects all of those by end of alpha-shift."

Werther got up to leave, muttering, "Okay, what daddy wants, daddy gets."

"Werther," she barked. When he turned back to face her, she said, "You're new on *Lovelace*, so I'm gonna give you that. Everybody gets *one*. But only one. The next crack I hear like that and you'll be running ladders with me 'til one of us pukes." She let that warning sink in before adding, "And, Chief? It won't be me."

He swallowed and nodded, answering, "Yes, sir," and quickly left.

Kennedy walked in immediately after him, asking, "You ready for lunch?"

"You bet." She nodded towards where Werther had retreated. "I suppose you heard that?"

"Yeah, I'm sorry. I thought Werther seemed like a good recruit for us."

"Oh, he'll probably work out fine." Decker swallowed the last of her coffee as she got Kay's bowl down from the shelf and filled it with food from the container next to it. "You know *everybody* has to make a big deal outta it at first. I don't know why." She gave the dog his lunch and instructed him to stay in her office. As the two Tactical Leads started towards the mess hall, Decker continued her rant. "It's not like I'm the only officer in the UDC with a parent up the chain. Considering how many goddamn legacies they admit at The Rock, it's downright common."

"Sure is – but you're the only one who serves directly under them."

"That's not true. Diego reported up to Admiral Romero for a tour or two and Liz Delaney posted into the *Caldwell,* which, strictly speaking is under her father's—"

"Yeah, but come on. You don't report *up to* Ricci – he's your captain. And adding fuel to the fire, no other legacy officer rooms with their parents. You think that bit of news doesn't get around? Think again."

She started at that piece of information. "*None* of the others room with their parents?"

"Yeah, why?"

"That's kind of weird." Decker shook her head. "Must be another *N'daa* thing I'll never understand." After a second of thought she exclaimed, "Hey! I'll bet that's why he—" She stopped and faced Con who paused as well. "When we were packing up to ship out, Ricci asked me if I was happy living there with him. If I maybe wanted to get my own quarters. I thought it was strange of him to ask, after all this time."

"Didn't you two discuss that when he asked you to move in?"

Deck re-started their forward progress, explaining, "No...we never *really* discussed me moving in at all."

"Uh, so what happened?" Con prodded with a chuckle. "Did you and Kay just show up one day with all your gear?"

"*No,*" Naiche countered. "After the Pakarahova mission, Ricci *insisted* on me staying there while my grandmother was...." Deck put a hand to her mouth as a sudden realization hit. "Oh shit. Maybe he only meant I should stay there *while the family was visiting.*"

"It's been well over a year. He woulda said something before now."

"Not Captain Contrition. He's probably been gritting it out all this time, because he thought he *owed* it to me." Naiche brushed a hand over her crown of braids and recalled, "And then there was how *weird* he acted about staying over at Stein's last time...." With a sigh, she admitted, "Looks like I've been cramping his style."

"What're you gonna do?"

"I guess when we get back to Uniterrae, I'll have to get my own quarters."

Con patted her on the shoulder. "It won't be so bad. You're a first lieutenant – you qualify for private quarters now."

"Do any of them have the full kitchen and a private balcony?"

Kennedy gave a short bark of laughter at her question. "No way. Sorry to inform you, Deck, but you've been living the high-life in captain's quarters. You're gonna have to come back down to earth."

After getting their food, they searched for a table and saw Aqila waving at them. She was sitting with one of her science officers, Second Lieutenant Blythe Brodie, and Tal Jacoway. Bly attempted introductions but Deck waved her off. "Yeah, it's okay. We've all met."

Tal looked back and forth between Aqila and Con, confirming, "You two are married, isn't that right?"

"Over a year, now," Aqila, answered, beaming at her husband.

"I also understand that Commander Sasaki is engaged to someone on board?"

"Yeah, Norm Avery in Engineering," Bly said. "And Commander Ramsey is married to Maisie Collins, the head chef."

Tal turned to Decker, observing, with an arch smile, "Wow, *Lovelace* is just one big happy *family*, huh?"

After the scene with Werther, Naiche was in no mood for Jacoway's sly digs. "Well, two out of three ain't bad."

"Not happy?"

"Not presently."

"Uh-huh." Tal stood up, saying, "Thank you for the company, Aqila and Bly, but I have to be getting back to work."

As soon as he was out of earshot, Deck quipped, "I hope it was something I said."

"Why? What's your problem with him?" Aqila asked.

"Ah, we knew each other at The Rock."

"You didn't get along?"

"Let's just say, we are not each other's favorite person."

"It's not what you're thinking," Con assured his wife. "They're not secretly related. It's a Chiricahua thing – we wouldn't understand."

Bly leaned in and offered up a mischievous grin. "You sure it's not a case of belligerent sexual tension?"

"Belligerent what?"

"Sexual tension. It just means that your animosity might stem from the fact that you're actually attracted to him but don't wanta admit it."

"I admit it."

Con's fork paused mid-way to his mouth and he stared at Naiche. "You do?"

"Of course, I do! Look at him. What male-attracted person in their right mind wouldn't be?"

With a huff of laughter, Lateef asked, "Did you ever think of doing...." She threw out a hand. "...*something* about it?"

"Oh, I *tried*, back before I figured out what a jackass he is."

Brodie blurted, "He turned you down!?"

Deck nodded, musing, "Inexplicable, isn't it?"

"Maybe he was seeing someone," Aqila suggested.

"Nah, the truth is that he just doesn't go in for casual sex."

"Oh, here we go," Bly muttered.

"While I don't go in for any other kind." Deck looked up and shook a warning finger at her companions. "That was *not* an invitation for the happily-marrieds to start lecturing me."

"It's not a *lecture*," Aqila objected. "I just think you're missing out on the possibility of—"

"Bly, you seem to be done eating. Would you pass your boss your hand-held?"

"What for?"

"I think she needs to look up the definition of the word 'lecture'."

***

Late that night, Decker was playing one-on-one basketball with Ricci in the deserted gym. It had become their habit to kill time that way when they were both suffering from one of their frequent bouts of insomnia. They were playing the game H-O-R-S-E, which Ricci had taught her. It was his turn and he called his shot. "Bank swish!" He released the ball and they both watched it bounce off the backboard and into the basket without touching rim.

"Damn," Decker said with grudging admiration. She looked at Ricci accusingly. "You know that's my worst shot."

"Of course, I do. I'm out for blood here."

"Well, I'm still two letters up on you," Naiche announced as she took the shot. It went in but hit the rim on the way. "One letter up."

Ricci then called a jump shot, which he missed. The ball bounced over towards Kayatennae who, as usual, was a keen spectator; he nosed it towards Naiche. She picked it up and dribbled for a few seconds while Matt said, "You never did tell me what's keeping you up tonight."

Decker called a 'straight in' which she proceeded to make. "Same thing that's got you up, I'm sure."

"You had a leg cramp, too?" Ricci teased, as he replicated her success with the shot.

"No. I'm up because I'm worried about the mission. And I'm betting you are, too."

Rather than denying her assertion, Ricci asked, "What's got you so worried?"

Decker called and made another shot as she explained, "Con and I listened to those transmissions from the *Daleko* about fifty-times today. We can't make any sense of what went on there. If there was an attack – it's not like anything we've seen before."

"Yeah, I know."

They played in silence for a short while, only broken by their called shots and the sounds of the game. After her next missed shot, Decker paused and asked, "You knew Captain Bastié, didn't you?"

"Yeah, I *know* Jackie. She's an old friend." Ricci ran a hand through his sweaty hair, protesting, "God, are we talking about her in the past-tense already?"

"Honestly? A lot of people are."

"What people?" Ricci had regained control of the game and called another bank swish. "These mysterious pilots at this mysterious bar that you mysteriously started frequenting?"

"Okay, you got me," Decker laughed, as she once again missed the shot. She took a deep breath. "I knew all about the QNS because I...." In a rush, she admitted, "I secretly took the simulator out for some test runs. A lotta test runs."

"Goddamn it, Naiche!" Ricci tucked the ball under his arm and turned to her, red-raced. "What possessed you to make such a reckless, bone-headed move? Haven't we talked about this kind of thing?" He angrily called a 'straight in' and missed, rattling the backboard with the force of his shot. Matt faced her again, insisting, "*Everything* about that ship was off-limits to you! Think how it would have looked if someone had caught you."

Rather than admitting that was exactly what had happened, she wheedled, "Ah, come on. You're a pilot. Deep down, don't you wanta take it for a spin? Even if it's just virtually?"

Ricci glared at her. "What I *want* is to see you use your abundant talents for something besides making trouble for yourself."

"Hey, we're on the *Lovelace,* now. You can't give me the 'father lecture' – you can only give me the 'captain lecture.'"

"Okay, as your captain, I'm *ordering* you to stay away from that fucking ship. If you so much as *breathe on it* in an unauthorized manner, I'm going to write you up *and* confine you to quarters." While Naiche called and hit a three-pointer, Ricci muttered, "And when we get home, I'm gonna ground you."

"Yeah, about that," Deck ventured, hoping that some good news would jolt her father out of his ire. "I decided that I *am* gonna move into my own quarters when we get back to Uniterrae."

After matching her three-pointer, he said, "You know I'm not actually gonna ground you – don't you? That was just a bad joke. I don't think of you as—"

"I know, but it's high time I got my own quarters, right?" She called and missed a set shot.

His mood hadn't lightened at all and he glumly replied, "Right, sure."

Recognizing that her irresponsible behavior was at the root of his irritation, Decker said, "And I'm sorry about hijacking the simulator. You're right. It was a...a bone-headed move."

Ricci nodded and offered a tight smile. "Thanks for telling me, anyway."

Still looking to get them back to their jovial rapport, Naiche bantered, "Your ball, Captain. But you're at H-O-R-S. One miss and I win."

"So are you. And I'm not gonna miss. You are," he declared. "One-handed reverse lay-up." Naiche groaned at the mention of another troublesome shot for her but was glad to see him swing back into the flow of the game.

Chapter 4

The Unknown Infinite

"The known is finite, the unknown infinite; intellectually we stand on an islet in the midst of an illimitable ocean of inexplicability." Thomas Henry Huxley

Two weeks into their journey to the Expanse, Communications officer Leticia Evans turned to Decker, who was the ranking officer on the bridge that afternoon, announcing, "Lieutenant Decker, I'm receiving a distress signal."

"From a ship?" asked Deck, scanning the Tactical console for signs of any vessels in their vicinity. There was nothing obvious.

"I'm not sure where it's coming from, sir, but its U-dec signature indicates it's from...." Lieutenant Evans paused, obviously rechecking her findings. "Yes, it's from the *Daleko.*"

"Transfer that to my station!"

Ten minutes later, Decker's station was surrounded by the entire command staff plus Charani and Jacoway. Lateef was leaning over, reading the data. "This doesn't make any sense." She turned to Ricci, saying, "It *is* from the *Daleko*'s mobile emitter but still...."

"It didn't fly away on its own," Decker finished.

Jacoway advised, "You should assess the signal for amplitude interference."

With poorly disguised irritation, Decker said, "Yeah, before I called everybody up here, I probably should've done that. And confirmed the U-dec signature against the database, done a vector field analysis, performed an LPS wave confirmation, run a Com-sat self-diagnostic and—"

"Decker." That one word from Ricci was sufficient – she knew better than to continue her sardonic litany.

"Could something be...mimicking the signal? Or rather someone?" Kennedy asked.

"Somebody setting a trap for us?" suggested Decker.

"Exactly what I was thinking."

"Since we can't figure out where it's coming from, it's not a *very good* trap, is it?" Lindstrom observed. "Unless they're assuming we can't resist a challenge."

Lateef said, "I'm going to perform a triangulation scan and see if we can pin-point the signal source. It's not a ship or a planet but it's got to be *something*."

After twenty-minutes of intense activity on Lateef's part, and anxious waiting on everyone else's, Aqila had an answer. "I think it's a small...satellite? Maybe? Or a drone of some sort."

"Is it stationary?" Ricci asked.

"No, it's on the move. Just not very quickly."

"Okay, let's catch up with it," directed Ricci. "Petrović, lock onto Lateef's scan and follow it to its source."

"Aye, Captain."

They quickly moved to the spot that their sensors indicated but couldn't get a visual on the source until it was greatly magnified. It turned out to be a drum-shaped object, metallic brown, about four meters in diameter, one meter tall.

His words slow with disbelief, Ricci asked, "What the hell is that?"

Lateef stopped gawking long enough to answer, "I have no idea."

"No weapons or defensive capability whatsoever detected," Kennedy reported. "But there might be a life sign?"

"What!?"

"Well...maybe? My sensors indicate a forty-seven-point three percent chance that what I'm reading is some kind of life sign."

"Looking at the size of that thing, smart money says no," Lindstrom declared.

"There's one good way to find out," Ricci said. "And that's to just bring it inside." He turned to the chief engineer. "Ramsey, can we use the Bessel hook to pull that thing in?"

"No problem, Captain," she laughed. "Hauling in a snare drum will hardly strain the engines."

However, the "snare drum" proved more troublesome than they'd anticipated. As soon as the virtual scoop got near it, the object darted away.

"Captain," Evans said. "I'm receiving a transmission from the...thing."

"Put it through."

Suddenly the bridge was filled with an ear-splitting, pain-inducing, screeching noise. Everyone immediately covered their ears in an attempt to hide from the anguish.

"Well...it had a weapon, after all," Ricci complained, as the din faded away.

"Sounded like Nik when I use up all of the hot water," Charani said to Ricci. Decker dared not look at them at that moment, for she would have been tempted to announce to the bridge that her own housemate did much the same.

Sasaki countered, "I don't think that was an attack." He was busily checking the data on his screen. "I believe that was an attempt at communication."

"If that's its hello," said Lindstrom. "I sure don't want to hear its 'go to hell'."

"Can you make anything of the transmission, Sasaki?" Ricci asked.

"Not as of yet, though the KD715 algorithm might—" Jeff's pronouncement was interrupted by another ear-splitting volley.

"Evans, cut all transmissions from that damn thing!" Ricci bellowed. "Ramsey, just bring it in here. We'll keep it in isolation while Sasaki's working on a translation."

Engineering made another attempt to pull in the small vessel but it suddenly fired an afterburner and zoomed out of sight.

Ricci thundered, "Where the hell did it go?"

"I'm scanning," Decker announced. "Nothing." She looked at Kennedy and Lateef. "I think it cut the signal on the distress beacon. We need to widen sensor scans."

The crew waited anxiously for a half an hour while Tactical and Scientific deployed all long-range sensors. Finally, Lateef announced, "I got it! It looks like it headed to the nearest star system, Captain."

"Okay, Petrović, follow that snare drum."

***

The *Lovelace* was circling a small watery moon where they had traced the location of the object, which had been emitting the *Daleko's* distress signal.

"What's the surface like?" Ricci asked

"Safe for human life," Lateef read off the results of her scans. "No signs of technology...no energy signatures of any kind. Mainly an aquatic world, no oceans though, rivers and lakes predominate while the land formations are covered with massive vegetation."

Charani asked, "Massive vegetation? Better known as trees?"

"If trees grew to be six or seven-hundred meters tall." There was a general murmur of amazement from the crew, and a low whistle from Con while Aqila continued, "And they're pretty close together. On the average, six to twelve meters apart. Even closer than that in the interiors."

"No wonder it's hiding down there," Kennedy said. "We'll have to stay above them."

Decker said, "As small as that thing is though...."

"Yeah, we might need to chase it through the vegetation. Could we pilot a shuttle around those things, you think?"

"I can," Jacoway volunteered.

"Not knowing what we're up against down there," Decker stated firmly, "a Tactical pilot is called for in this—"

"I think the *best* pilot is called for in this case. And that would be me." Tal looked around. "Can anyone dispute that?"

"What I can dispute is your—"

"That's strange," Ricci broke in. He asked Lindstrom, "When did we start assigning mission personnel by volunteer? I must have missed that memo." Matt turned to his chief tactical officer, ordering, "Kennedy, assemble your team."

"Yes, sir," Con answered.

Tal's shoulders slumped in dejection, figuring that he would be shunted aside. He refused to look at Decker, knowing she'd be grinning at him in derision.

"Ensure that you consider *all* your options, though, Commander."

At that firm piece of "advice" from the captain, Jacoway couldn't resist a glance at the Tactical leads. There seemed to be an unspoken conversation going on between the two. When Decker gave the tiniest nod of her head, Kennedy turned to Tal and said, "Okay, Jacoway, you're the pilot." He then said to Naiche, "Deck, we might end up on foot so I'll need you and Kay. We'll bring Kapoor, too." Sasaki and Lateef were chosen to round out the team.

***

The terrain proved to be as challenging as they'd feared but Tal was piloting expertly through the thicket of tree-like structures. They had turned out to be huge, thickly branched spikes covered in what looked like heavy purple moss. He was pushing the shuttle at full speed, following the signal of their scanner, which waxed and waned slightly for unknown reasons. While the shuttle had to dodge around the trees, the small ship wasn't so hampered but Jacoway was managing to stay in close pursuit.

Kennedy said, "We've got to catch up to that fucking thing before it gets closer to the interior." Their scans showed that in the center of the "forest" the trees were so thick that the shuttle wouldn't be able to fit – not even in the hands of a pilot like Jacoway.

"The question is," Decker said, "why isn't it there already?"

Though Con nodded in agreement with her assessment, Lateef asked, "What do you mean?"

"It had the jump on us by about an hour – it coulda been there already. It's moving like it knows the lay of the land, so why didn't it head right for the interior?"

"Maybe it didn't think we'd follow," Corpsman Priya Kapoor suggested.

"Maybe," Decker answered. "I can't help but feel like we're missing something."

Twenty minutes later, Jacoway looked at his screen and saw that he was closing in; the same thrill of victory that he remembered from his fighter pilot days, coursed through his veins. They were reaching a clearing, which would be perfect for capture of the little vessel. He crowed, "We've got it! It's dead ahead and we'll over-take it in a second."

Kennedy turned to Decker in the co-pilot seat and said, "Get the grapnel ready."

Another turn and twist around some trees and the ship was in sight. However, as they drew near, it was evident that it had landed. Seeming dead, it was emitting no sounds or energy signals. Jacoway landed the shuttle right behind it.

"Did it lose power?" Deck asked.

"We better check it out," Kennedy responded. He and Decker moved quickly to the rear and grabbed particle rifles. He tossed one to the corpsman, saying, "Kapoor, you're with us." Con paused at the hatch and ordered, "The rest of you stay here. This could be a trap."

As they watched the Tactical team advance cautiously towards the tiny ship, Tal asked, "Why does the dog get to go? What's he going to do?"

Lateef said, "He's not a dog, he's a corpsman, third class." Jacoway wasn't sure how that answered his question but he let it go.

After some fruitless scanning, examination, and prodding at the ship, Kennedy told the remaining crew over the comm that it was safe to approach. "It seems empty," he announced as they gathered around it. "That life-sign, if that's what it was, is gone."

"What is that noise?" Aqila asked, referring to a constant whirring sound in the air.

"Sounds like a *tzi-ditindi*," Decker said.

"A what?"

She hesitated, as if trying to figure out how to describe it. "It's a...musical instrument, I guess you'd say. *Ndee...uh,*

Apache use it in some of our spiritual rites. It has a weighted wooden airfoil attached to a long cord. You whirl it in a big circle and it makes the sound. Carries quite a ways, too."

"Oh, a Groaning Stick," Jacoway said. "That's what the Navajo call it."

"Are you Navajo?" Sasaki asked.

"He's *everything*," Decker interjected.

"My people are part of the One Nation Collective and draw our strength from multiple cultural traditions," Tal insisted.

Priya was looking around her. "I don't know what you call that noise but I think the trees are making it."

Kennedy looked up at the massive, purple structures. "I think you're right." He raised his eyebrows at Aqila. "Spooky." Con clapped his hands and said, "Okay, Deck, Kapoor, get to work! Scan the clearing for signs of our...prey," he finished with a laugh.

Decker started out on a brisk trot, with Kay sniffing eagerly behind her. The other two Tactical crewmembers went off in opposing directions. After a few minutes they came back empty-handed. Decker said, "Absolutely *nothing* coming up on the scanner." She kicked at the rocky soil, saying, "I didn't see anything that looked like tracks, either. Not that I know what kind of tracks I'm looking for."

"Tiny ones," Lateef suggested.

"Maybe it can float," Tal said.

"Maybe it can float, maybe it's invisible, maybe this is all a fever dream due to my reckless ways finally catching up with me," Decker said.

"Anything more useful to offer?" Con laughed.

"If we could pry that damn ship open, at least Kay could get a sniff of the interior and know what he's supposed to be tracking," she replied, pointing at the eager dog.

"Okay, let's give it a shot." The rest of the crew lounged against the shuttle, watching Con and Deck work to open the vessel. Finally, Kennedy straightened up and wiped a hand across his forehead. "This isn't getting us anywhere."

Kapoor pointed her rifle at it, saying, "What about blasting it open?"

Decker shook her head and motioned for Kapoor to lower her weapon. "No. Number one, that wouldn't leave much for Kay to sniff, and number two, the emitter could still be aboard that thing for all we know."

Kennedy looked around and said, "Let's at least perform a more thorough search of the wider area before we give up. We'll split into two teams. Deck, I'll lead one and you lead the other."

"Sounds good." The two tactical leads conferred over their scanners, which were displaying maps of their current location. Jacoway and Sasaki were chatting near the edge of the tree line when the sound of Decker saying, "Okay, I'll take Sasaki and the peacock with me," drifted over to them.

Jacoway bristled at that nickname, while Sasaki asked, "Are you '*the peacock*'?"

"Yeah," Tal grumbled. "That's what Micro-craft pilots called fighter pilots during the war."

"And you called them...?" asked Sasaki, knowingly.

In a sheepish tone, Jacoway admitted, "Pigeons."

Jeff laughed. "It's inevitable. Even with us all speaking Standish, individual groups will *always* develop their own 'language.'"

"I guess. Thank god Terrans aren't still dealing with the 'Tower of Babel', though. We can all *understand* each other." He shot a glare over at where Con and Deck were still conferring. "More or less."

"That's true, we've gained a lot, but as a linguist I still worry about what we've lost."

Jacoway turned to Jeff with a wry smile. "Don't worry, Sasaki, you still have plenty of work translating alien languages. You're not obsolete, yet."

"That's not my concern. The language we speak affects how we experience the world, how we understand basic concepts such as color, time, direction – even life and death. What ways of human understanding have we already lost?"

"You're not saying you want us all to go back to speaking *different languages,* are you? The member states of Uniterrae have enough trouble getting along as it is!"

"No, I'm just saying—"

At that moment Decker strode up to them announcing, "Okay, Commanders, you're with me; we're covering the southwest." She pointed to their designated area on her scanner map.

"Looking for?" Sasaki asked.

Decker exhaled, answering, "Whatever we can find."

As they started off, with Kay bounding ahead, Tal asked, "Tell me, Decker, do you call *your father* a 'peacock'?"

"No, sir. I call the captain, 'Captain Ricci'. I would have thought you'd have heard that." She turned back to him, saying, "I am a trained medic, if your hearing needs—" Just then Kay yelped and they all turned to see him sprawled on the ground in front of a couple of slanted trees. Decker ran over to him. "Hey, buddy, what happened?"

The dog jumped up and shook off, apparently unharmed. The group started walking again, with Naiche in the lead this time. "Were you so engrossed in sniffing that you walked right into a tree? You need to watch where you're—" The open space Decker had been headed towards seemed to have disappeared, now filled with two trees entwined together. "That's...weird." She backed up and headed for the next space only to find herself similarly blocked. The trio watched open-mouthed as all of the trees in sight could be seen twining together, forming an impenetrable barrier.

Tal asked, "These trees...they're moving. Am I seeing this right?"

"They're not just moving," Sasaki, answered. "They're actively blocking our way."

"What the hell does this mean?" Jacoway demanded.

"It means that these so-called trees both *talk and move*." Decker said, "It also means the chance of this being a fever dream of mine just went way, way up."

Chapter 5

Veil After Veil

"Veil after veil of thin dusky gauze is lifted, and by degrees the forms and colours of things are restored to them, and we watch the dawn remaking the world in its antique pattern." – Oscar Wilde, The Picture of Dorian Gray

With a huff of bitter amusement, Captain Ricci looked at the crew gathered in the war room, and said, "So, basically we can't get at this thing because it's being protected by a forest of Ents." He ran a hand though his hair, while muttering, "Fuck my life."

"Ants?" Decker asked.

"It's from an ancient *N'daa* book. I'll explain later."

"Are we even sure that the...*pilot* is still around?" Charani asked.

"Yes," Lateef answered. "Our deep-scan sensors have confirmed that both the original life sign and the mobile emitter are in the forest interior."

"Why weren't we picking it up down on the surface?" Con asked.

"I'm not sure – but my theory is that those sounds that the trees make create a resonance that interferes with our scans at close range."

"Can't we simply blast our way through?" Lindstrom asked.

"Those trees just might be a sapient form of life," Aqila objected. "And they've made no hostile move towards us."

"They're harboring a known fugitive," Decker said.

"That's a stretch," Ramsey said. "Is it worth all of this trouble to get the emitter at this point?"

Charani explained, "If we can upload the data from that emitter, we'll know the coordinates of where it was first activated. Which is presumably where the *Daleko* is."

"Or rather was," Ramsey said. "It was either destroyed or they abandoned ship or they wouldn't have activated the mobile emitter in the first place. So, I can't help but wonder if Bastié and Pilecki are even still alive—"

"We owe it to them to find out!" Jacoway insisted.

Leaning forward in concern, Ramsey objected, "Even with our latest data about the Expanse?"

"What data?" Zache asked.

Ramsey looked from the captain and Lindstrom to Lateef. "They haven't been briefed?"

"I just saw the data, myself, twenty minutes ago, Carla," Aqila answered.

"What data?" Zache repeated.

Lateef launched into the explanation. "Astrophysics, with the help of Lieutenant Brodie, has been monitoring the status of the Expanse. Brodie's latest model indicates an increase in instability on an exponential scale. The anterior margin may already be shifting."

Ricci asked, "Do we know how long we have before it disappears from this sector?"

"If Bly's model is correct, then we have twelve days – at best."

"How far away are we, at our present location?" asked Charani.

"We're still four days away," Ramsey replied. "Even at top speed."

Zache said, "I guess that seals it, we can't waste any more time here." As several people directed surprised looks at her, she added, "Look, if there was any kind of guarantee that we could help Bastié and Pilecki, I'd say, yes, but as Ramsey just indicated—"

Kennedy insisted, "There's got to be a way into that forest! We haven't checked the entire perimeter."

Lindstrom said, "Do you know how much time that would take?"

"Do you?"

"More than we've got."

"Give us twenty-four hours. We'll take every crewmember that *Lovelace* can spare. We can bivouac there and rotate teams in and out – go round the clock."

Ricci said, "Okay, say you can find a way in, how are you going to locate the pilot? Scanners seem semi-reliable at best."

"Kay and I can find it," announced Decker.

"You said yourself he doesn't know what he's tracking," Tal objected.

"He could track the emitter. There's an equivalent one on the *Cerxai*, right? It's got to smell similar enough to the one from the *Daleko* to give Kay the scent."

Ricci leaned back in his chair, parsing all the many moving parts of the plan. "This sounds like a long shot...."

Kennedy pleaded, "Captain, give us the twenty-four hours. Like Jacoway said, we owe this to Bastié and Pilecki – it may be their only chance."

Decker added, "If it was *me* stranded somewhere, I'd hope the Corps would do as much." She looked around the room. "I bet deep down you all feel the same way."

"All right, Commander," Ricci said to Kennedy. "You've got those twenty-four hours." While Con was thanking him, Matt warned, "But no more than that."

***

Trooping through the darkness, the rocky path lit mainly by their wrist lights, Jacoway brought his tired, discouraged team back to camp. Their four-hour search for a break in the tree line had been unsuccessful. In the yellow-white glare of the camp-lights, he waited to talk to Decker who was pacing back and forth while conversing with Kennedy over their comm links. The chief tactical officer was running the main camp, located on the far side of the forest from them.

Tal could hear Kennedy gently scolding his second in command. "Deck, you've been up for almost twenty hours straight now. You need a nap. If we do find the entry point, you and Kay have to make the run to where—"

"I know that, but I also know I'm not gonna sleep, you know I'm not gonna sleep, so what's the point of this discussion? I'll make the run, come hell or high water." She massaged her right temple, asking, "Have we heard anything from Werther's team?

"Yes, he reported in a few minutes ago. Nothing on the north side. I'll let you know when Abello reports in."

"Okay. I'm sending a triple team out at dawn. That's our last chance. Decker out."

Deck turned to Jacoway. "Welcome back. I can tell by your face that asking how it went is unnecessary."

"Yeah, those trees have thrown up defenses tighter than The Rock's front gate."

"Okay, your team's on a three-hour break. Make sure everyone gets some chow and a nap. We're *all* heading out again at daybreak."

"Maybe you can find a spare rifle for me by then?"

She sighed wearily before answering, “Commander, I *told* you, you’re not weapons qualified.”

“Do you actually think I didn’t have training during the war?” When she gave him a skeptical look, he said, “Sure, it was only during Squad-entry but still....”

“So...six weeks of training, how many years ago? You are not *currently* qualified and—”

“How difficult is it to handle a particle rifle?”

Decker stood stock still for a moment and then snapped, “You want a rifle? Order me to give you one.”

“What?”

“You out-rank me. Give the order.”

Feeling like this was a ploy to make him look foolish, Tal nonetheless said, “Lieutenant Decker, I order you to give me a particle rifle.”

“Yes, sir.” Decker whipped her own rifle off her back and handed it to him, saying, “Here. Now it’s on you if something goes wrong and you hurt yourself or someone else.” She walked away stating, “I have enough deaths on my conscience.”

She stalked over to a boulder near the river and sat down with her back to him. Tal swore Kayatennae gave him an accusing look before following Decker to her perch. Jacoway thought it best if he and Decker both took some time to cool down. He called over, “Okay, I’m going to grab a bite to eat,” and ducked into the tent where stacks of meal packs, hot coffee, and water tubes could be found.

Forty minutes later he found her still sitting on the boulder but this time she was busily entering data into her hand-held. He walked over and handed her a cup of coffee. “Thought you could use this.”

“Thanks.” Naiche took a sip, saying, “You mean because I’m so cranky?”

“That...and from what little I’ve seen, you drink about a gallon a day.” She gave a weary smile and nodded. Tal took the opportunity to lean the rifle on the small boulder next to the one she was sitting on. "Turns out I don't need this thing after all. Wouldn't wanta kill someone with my incompetence."

Exhaling softly, Decker said, "Yeah, I guess I should've been more *diplomatic* about that." She rolled her eyes, adding, "Especially considering my *heritage*." Naiche laid her hand-held down beside her and stretched her back. "Looks like I still don't have my mother's aptitude for it."

"No, that whole thing was on me. You were just following regs." Tal was debating whether to add, 'for once in your life' but was distracted by a splash from the river. "What was that?"

"There's some kind of large creatures living in these waters."

"Really?"

"Yeah, Aqila says they're the most abundant form of life on this moon but I've never actually seen one. I've heard them plenty, though – so has Kay."

Jacoway slid up onto the boulder beside Decker and glanced down at her hand-held that lay between them. "You write your reports while still on mission? I'm seriously impressed."

Decker held it up, giving him a better look. "I was working on my monthly message to my Great-Aunt Loza. I don't write my reports in Chiricahua."

He studied the unfamiliar text, saying, "Yeah, of course not." Jacoway took the opportunity to ask a question that had been nagging at him. "Which language do you think in? Generally, I mean."

"What?"

"To use the QNS simulator you must be able to think in Standish but—"

"Oh, I see. Yeah, I can think in Standish now. I usually do when I'm speaking it." Deck gulped some coffee before adding, “In case you haven’t noticed, my Standish is pretty good these days.”

“What does that mean?”

She studied him for a second before saying, “Back at The Rock – you said my terrible Standish made me a useless study partner for you. I heard you telling Bian Nguyen in the library.”

Jacoway’s stomach plummeted and he was momentarily speechless, as the precise conversation Decker was talking about came flooding back to him. “You *heard* that?”

“Yeah, my Standish might have been awful but my hearing has always been aces.”

Some long-missing puzzle pieces fell into place for him. “*That’s* why you asked to be re-assigned.”

“Yes.”

He thought for a second and ventured, “And that’s why you pulled that ‘Do you speak Choctaw?’ stunt. Isn’t it?”

She had the grace to look slightly abashed as she answered, “Yeah, that’s why.” Deck exhaled a breathy sigh and then offered, “Sorry ‘bout that. It was...a pretty juvenile reaction.”

Tal shrugged, saying, “Well, that’s not so surprising – considering how young you were at the time.” He silently considered whether this was an appropriate occasion to dissect their mutual misunderstandings. Coming down on the side of, “no time like the present,” Jacoway said, “The truth of the matter is—”

He was interrupted when Kay started barking furiously and ran towards the riverbanks where a faint splashing could be heard. “Yes, I know about those creatures, Kay. We’ve been over this.” The excited dog ran back towards the adjacent boulder and barked even more frantically. “Fuck! The rifle is gone,” Decker exclaimed.

Before Tal knew what to think, Naiche was stripping off her jacket and pulling off her boots.

"What're you doing?!"

"I'm going after it. We can't let some animal get its hands on a particle rifle."

"That water is freezing!"

"I'll be fine."

"Decker, this is too danger—"

She paid him no heed, handing him the moonstone necklace from around her neck, saying, "Here, hang on to this for me." Deck ran to the river calling back, "Contact Kennedy immediately – let him know what's happened and where I've gone." As she dove in, Naiche yelled at the dog who was running after her, "No, Kay – stay!" And then she disappeared from sight.

"Decker!" It was too late. Tal could no longer see her, only hear the slight splashing of her swim strokes.

***

Shortly after he'd contacted the other camp, Tal was watching the shuttle land. Kennedy and Lateef jumped out and rushed over to him; breathlessly, Kennedy asked, "Any sign of her yet?"

"No, nothing yet."

"Fuck! She's been under for over twenty minutes!"

Lateef clasped him on the arm. "Con, we don't know that she hasn't surfaced somewhere else."

"She isn't responding to her comm."

"That doesn't mean she's—"

"I know, I know." Kennedy ran a hand over his short, tightly coiled hair, lamenting, "We can't even pick her up on our scans."

"*Lovelace* might be able to!" Aqila exclaimed. "Let them try."

"Right! Great idea." Kennedy turned to the Tactical lieutenant, Evelyn Bayer, standing behind him and said, "Bayer, what time is it on *Lovelace* now?"

She consulted her hand-held and said, "It's 0450 hours, sir."

Kennedy tapped the comm link in his ear. "Kennedy to *Lovelace*, come in *Lovelace*."

"*Lovelace* here."

"Hey, Mikkelsen, patch me over to whoever's at the Science console right now."

"It's Gardenalia, sir," answered Mikkelsen, performing the transfer.

"Gardenalia, I need you to do a deep-scan sweep of the surface for Lieutenant Decker's comm signal."

"Yes, sir. Scanning." They waited and a moment later, the ensign reported, "I have it, Commander."

"Good. Now scan the immediate vicinity of that signal for human life-signs."

"I'm on it."

Tal could recognize the tension in the set of Kennedy's jaw and the stock stillness of his body. Lateef was waiting with her eyes closed and hands tightly clenched.

Finally, Gardenalia said, "Commander, I'm reading one human life sign in that vicinity. She's about four-hundred meters from your present location."

After expelling a long breath, Con asked, "Exactly how many non-human—"

"Kennedy!" Ricci's voice broke into the conversation. "What's going on down there?"

Con put a hand to the comm-link in his ear and turned to Lateef. "Son of a bitch! Doesn't *either* of them ever sleep?!"

"Kennedy! Do you copy?"

After quickly uncovering his comm, Con answered, "Yes, Captain, I copy. We're...uh...trying to locate Decker, sir."

"She's missing?"

"Yes, sir. It looks like one of the indigenous species here made off with a particle rifle and Decker chased after it. And uh, she hasn't reported back yet, and isn't responding to hails."

"How long has she been gone?"

"A little over twenty-minutes."

"Is Kay with her?"

"No, sir, he's here with me."

"Then send him after her."

"The thing is...she went into the river after one of those water creatures. But we know she's alive, Captain, since we have her—"

"She's been under the water for twenty minutes?!"

"We believe she must have surfaced...somewhere."

"But she's not checking in or responding to hails?"

"No, but we have a fix on her now and we're heading out to—"

"I'm on my way dow—" Ricci stopped abruptly, evidently wrestling with his dual role of captain and father. There was a moment of tense silence before captain won out. "I'll be standing by. Let me know immediately as soon as there's any word on her. Ricci out."

Kennedy briefly rubbed his forehead and hit his comm again. "Gardenalia, still there?"

"Yes, sir."

"Send a map of Decker's location to my scanner." Kennedy studied the transmission, muttering, "Okay, that would put her directly in the middle of the water so there must be a rock or something...." He looked over his shoulder and said, "Bayer, you and Kapoor deploy two skiffs. Let's get out there and see what's up with Decker."

"Yes, sir."

Kennedy nodded at Kayatennae who was prancing anxiously. "Yes, you can come, too."

Tal waited on shore with Aqila for thirty anxious minutes. When the two boats breeched the circle of light from the camp, it was immediately evident that Decker wasn't in either of them.

Trailed by a dejected Kay, a grim-faced Kennedy climbed ashore and embraced his wife. "There's no sign of her. I even re-confirmed her location with Gardenalia. We were supposedly right on top of her and – nothing."

"Do you think the sensors are malfunctioning – or inaccurate?" Jacoway asked.

"Of course, they are," Kennedy snapped. "Useless fucking—" He took a deep breath and said, "I'm sorry. It's just..." He shook his head and didn't finish. Then he looked at Aqila, saying, "I gotta contact the captain. He's probably paced a trench right into the deck by now."

Before Con could tune his comm-link to Ricci, Decker's voice crackled out of it. "Decker to Kennedy. Con, do you copy?"

"Naiche Decker! Where *in the hell* are you?"

"I'm in an underwater, umm, cavern I guess you'd say. There's plenty of air but— No, he's not the one you want. Yes, I will be getting them down here. Just give me a second!"

"Deck, who are you talking to?"

"The Faleena Nock— Well, I'm sorry, I'm not a linguist. The Fwalayna Noctay," she corrected very deliberately.

"Deck, are you sure you're okay?"

"Yeah, I'm soaking wet, I'm freezing, and I've been cross-examined by some self-important walrus-squids but other than that – I'm fine."

"Some self-important *what*!?"

"Look, just tell Ricci that he has to get down here pronto. The reigning residents of this planet want to talk to him. I'm bringing them to the auxiliary camp. Make sure Sasaki is available. Oh, better get Lindstrom, too. They feel slighted enough. Don't want to add to it."

"Who feels slighted?"

"The Fwaleenie whatevers, the aliens I'm— Yes, I know technically *we're* the aliens here, would you just let me talk to my people so I can get you what you want!"

"Deck, I got it. I'll make sure everyone's here to greet...them. Do you need anything else?"

"Yeah, a dry uniform and a vat of hot coffee. Decker out."

Chapter 6

Far More Powerful

"Never underestimate the value of knowing another's language. It can be far more powerful than swords and arrows..." Melina Marchetta, Finnikin of the Rock

Though he had graciously and humbly accepted it, Ricci had never felt that he'd truly deserved the Founder's Medal of Honor. He finally believed he'd earned it that day when he refrained from gathering a thermal blanket-draped Naiche into a hug and instead simply greeted her, "Lieutenant Decker, good to see you're well. What's the situation?"

Minutes later, in the faint light of a new dawn, he, Lindstrom, Sasaki, and Decker were standing on the banks of the river, which he'd been informed was called the Pyallenera, on this world known to the inhabitants as Vwalloon. The inhabitants themselves were called the Fwalayna Noctay. The aquatic species had large bulbous bodies, covered in smooth shark-like skin, which was the same gray-green color as the river itself. They looked to weigh at least 150 kilos, had five long, feathery, tentacle-like appendages and a large hollow opening in their heads. The opening was covered by a thin, netted membrane. When the membrane vibrated, it made sounds that were capable of a credible imitation of human speech. Their syntax, though, left a lot to be desired.

The leader of this group identified themself as Vertee Pwawheel. Splashing one of its appendages in the river, Vertee said to Ricci, "Seat at we on drink. Bigger for we."

Ricci looked at Sasaki for help. "Captain, I believe they're asking us to join them in the water."

Another Fwalayna spoke to Vertee in their own language. Decker whispered, "I think that one is their Sasaki."

Vertee pointed at some rocks jutting out of the river, repeating, "Seat at we on water. Bigger for we."

"I'm sorry," Ricci answered, "But we're not...equipped to stay in the water for long. It could hurt us."

The alien linguist once again helped out with a translation and then Vertee waved at Decker. "No hurt at Dek-ler. Dek-ler joined water."

"That's because she's crazy," Lindstrom answered.

While Ricci was glaring at his first officer, Vertee said to Decker, "Give at those your craw-zee. Have more you craw-zee?"

"Oh, I have plenty more where that came from."

Ricci interrupted this tangent, stating firmly, "I'm very sorry but it's not possible for us." He walked forward until the water was almost over his boots and motioned the others to follow suit. Lindstrom begrudgingly complied, as did the others, much more willingly. "Is this better?" he asked Vertee.

"Small bigger, Reeshey."

"Good, good. How is it you speak our language?"

"Now you at Vwalloon, Fwalayna take in your sounds. Take in sounds for times long and we speak too."

Sasaki said, "They've been listening to us and—"

"I got it, thanks."

Vertee asked, "Tear-rans not take in sounds from not-Tear-rans and speak too?"

Ricci admitted, "Umm, not as quickly, or as easily as you."

"This why you not make speakings at we, when now at Vwalloon?" The alien seemed to grow agitated, complaining, "Big Tear-rans," Vertee gestured at Ricci and Lindstrom, "not greet at Vertee Pwawheal." Vertee then pointed at Decker and Sasaki. "More small Tear-rans not greet at Vertee Pwawheal, too."

Sasaki spoke to his counter-part, called Cranelle Nadoosh, saying, "Yes, we are very sorry we didn't introduce ourselves. We meant no offense."

"Offense we have," Vertee insisted. "Is right all ale-lee-an ones greet at we when now at Vwalloon!"

"Yes, we acknowledge that we made a mistake," Ricci insisted. "We most certainly should have...attempted contact with you before setting up our camps." Matt knew there was little to be gained by trying to explain that the Terrans hadn't even recognized the Fwalayna Noctay as an intelligent species. "But we meant no harm to you or your world."

"You have no harm? For why bring well-pawns at Vwalloon?"

Ricci was lost for a moment but then Vertee gestured impatiently at the particle rifle that the Fwalayna Noctay had confiscated. "Oh, weapons. Yes, well, we didn't know what to expect here. The particle rifles are for protection and defense only."

Vertee needed Cranelle's help with that but finally turned back to Ricci asking, "For why Tear-rans now at Vwalloon?"

"We are tracking down a missing piece of our technology. We need it to help some of our people who are in trouble."

"Speak you at noise rock Dardanze bring now at Vwalloon?"

"Noise rock?" Ricci straightened up in surprise and said to Lindstrom, "The emitter! They know about the emitter!"

***

A short while later Charani was idly listening to Kennedy and Decker's conversation while watching Ricci and Lindstrom who, with Sasaki's assistance, were finishing up negotiations with the aquatic aliens. Decker had filled her CO in on the gist of the discussion between the Terrans and the Fwalayna and she and Con were presently discussing retrieval of the emitter. Zache was watching Matt work his undeniable charm on these strange beings, thinking, *Wow, Matteo Ricci, this is your life, now, huh?*

"I can do it," Decker insisted. "I've got enough adrenalin coursing through me to power the shuttle."

Kennedy said, "Besides your *obvious* exhaustion, you've been through *enough*, Deck. Anyone can follow Kay to that emitter." He gestured at the dog who was dozing at Naiche's feet.

Sasaki joined the discussion, interjecting, "Actually, we may not need to retrieve it, at all. It may very well be coming to us."

Charani, Kennedy, and Decker said in almost perfect unison, "What?"

The command team joined the discussion and Sasaki asked, "Is it settled, Captain?"

"Yes," Ricci answered. "The Fwalayna Noctay are going to explain our situation to the Whur-Whur-Whur."

"To the what now?" Charani asked

"That's what the trees are called."

"What was it again?"

"I'm not repeating it, Zache," Matt declared. "It's bad enough I'm going to have to put it in my report." He shook his head and continued, "Anyway, the hope is that the trees will in turn explain it to the Dardanze, that is, to the current holder of the emitter," he elaborated before anyone could ask. "Who will bring it to us."

They all looked towards the river where one of the Fwalayna Noctay had started to wave its tentacles about, producing a very close approximation of the sounds that the trees produced.

"And this Dardanze thing will just hand over the emitter, *why*?" Kennedy asked.

"Because, we've promised to give it back after we get the data off of it. To allow it to finish feeding."

"Wow, I guess I should have stayed for the whole conference," Deck mused. "Did you just say that it's feeding? *Off the emitter?*"

"Yes," Ricci said, very deliberately. "Apparently the thing that took the emitter ingests sound waves as its source of nourishment. It comes to this planet on a regular basis to feed off the tree resonances and in turn exudes...something they need. We didn't quite catch what. That's why the trees offered it sanctuary. From us 'Tear-rans' – who *maliciously* tried to capture it."

Lindstrom snarked, "Yes, this *all* could have been avoided if only we spoke air-attack siren."

"Come again, Commander?" Kennedy asked.

"We were scolded by the kraken back there, since the snare drum explained it all to us and *we just wouldn't listen.*" Lindstrom glanced back towards Vertee, stating, "They're quite arrogant for a species whose highest achievement is tentacle ventriloquism."

"That's not really true, sir," Decker objected. "They have quite a bit of technology in their underwater...um, town."

"Why didn't we pick it up on our scans?" Ricci asked.

"It seems to all be based on water flows. I guess that doesn't register as technology to our sensors." Decker warmed to her subject, expounding, "They captured recordings of us through etchings on rocks. It plays back when water flows over it. And their version of data screens appear as ripples on the surface of the water. I think they read them with their tentacles. It's all pretty cool."

"Really?" Ricci exclaimed. "Well, we'll be expecting a detailed report on all of that, Lieutenant. Once you've gotten some sleep and Medical has checked you over. Doctor Clemente is standing by on *Lovelace* to—" Matt was interrupted when Kayatennae woke suddenly and started barking at the tree line.

The *Lovelace* crew turned towards the source of the dog's agitation and saw that the tightly knit group of trees in front of them was slowly parting ways. To their amazement, a wheel-like object rolled out of the forest and approached them. The wheel resembled a cage of bone spokes with a luminous crimson gel-sac inside. Decker had to order a curious Kay to stay away from it. After a breathless moment of anticipation on the part of the Terrans, an opening formed in the side of the wheel and part of the sac extended from it, bearing the distress-signal emitter.

Kennedy reached down, took the emitter from the alien, and slotted it into his hand-scanner. "Umm...thanks. This won't take a minute." When the Dardanze answered in its own language, Kay ran away yelping while all of the humans immediately covered their ears.

Occupied as she was in protecting her hearing, Charani couldn't help but think, *Wow, Zache Charani, this is your life, now, huh?*

***

Later that day the senior staff was gathered in the war room listening to Lateef report on the data from the distress-signal emitter. "The emitter was initially activated on a small planet in the Antares star system, designation, Scorpii-d. At this range we can't tell much about the planet but we're within two day's travel of it."

"That's good," Kennedy said. "We can swing by on our way to the Expanse."

"Not really," Ramsey objected. "That's in the opposite direction."

Appearing slightly dejected at that bit of news, Zache asked, "Is it even likely that the *Daleko* ended up that far from the Expanse?"

"How likely is *anything* that we've experienced on this mission to date?" Ricci answered. "We only have this information because some water aliens convinced some trees and a red sac of goo to help us out. We only met those aliens because Lieutenant Decker is a strong swimmer and careless with her particle rifle."

Jacoway looked at Decker, waiting for her to call him out as the real culprit of that last part but she simply inclined her head as if acknowledging her mistake.

"Risking even, at best, four days delay though...." Charani mused. She asked Lateef, "What's the current status of the Expanse according to Brodie's model?"

"We're still seeing a steady decline in stability at present. It looks like the Expanse will disappear from this sector of the galaxy in ten days."

Ramsey asked, "Knowing that, is it worth the risk to go to Antares?"

"I don't think we have any choice. We cannot ignore a distress beacon. It would be unconscionable," Clemente said.

"The Corps directives are very clear on that point," Sasaki added.

Charani was shaking her head doubtfully. "Usually yes, but the way we've discovered this beacon is hardly the norm. It could very well be leading us away from our core mission."

"I don't think that changes our responsibilities regarding a distress signal one bit," Decker offered.

Jacoway asked, "What about our responsibilities to the *Burnell* and the *Meitner*? We could be condemning eighty people to a—"

Kennedy countered, "We don't know that we're condemning *anybody* at this point!"

"The probabilities sure point that way though," Ramsey argued. "We go to Antares and, at the very *best*, we'll reach the Expanse with less than two days to find those ships."

Ricci put his fingers in his mouth and gave a sharp whistle, putting an end to the main argument and several noisy side debates. "Enough! This is no longer a productive discussion."

"I do see both sides. How can we possibly decide?" Jacoway asked

"*We* don't," Ricci pronounced, pointing at the gold braid adorning his uniform wrists.

There was a murmur of "Yes, sir" and "Of course, Captain," from the group, though Jacoway and Charani shared a puzzled look.

"Gold braid decides," Lindstrom explained.

Tal nodded slightly, understanding the message but not the phrasing, so Ramsey clarified, "It's an old Command Operations saying." She pointed at the head of the table where Ricci was sitting. "Gold braid decides." She pointed opposite him to Lindstrom as she continued, "Silver braid confides," and finally she gestured around the table, finishing, "Silver stripes abide."

"Raw recruits have an addendum. White stripes die," Decker added, referring to the white stripes that were an ensign's insignia. A glare from Ricci was sufficient communication of his opinion about that supplement. She sighed apologetically, "I know, Captain. Not helpful. Sorry."

"So. Sir," Lindstrom said, "Tell us what the gold braid has decreed."

"We owe it to Bastié and Pilecki to check this out. They asked for our help. We have to at least make an attempt to give it. Not only that, but they may have information that the *Cerxai* needs to navigate the Expanse successfully. If all goes well, we should have sufficient time to rescue them – and the *Burnell* and *Meitner*."

"And so much has gone well for us so far!" Lindstrom said.

"I would take that to mean that we're due for some good luck, Commander," Ricci countered.

"Yes, Captain, of course," Lindstrom agreed. "Just *confiding* my concerns."

"Duly noted," Matt said. "Ramsey, lay in a course for the Antares system. Maximum speed."

"Aye, Captain."

"Meeting adjourned. Dismissed."

***

Jacoway managed to catch up with Decker as she and Kay were heading to her quarters. "Decker!" he called. When she turned towards him, Tal said, "Hey, I don't want to keep you, anyone can see you're absolutely fried, but I did want to say thanks – for not calling me out in the meeting about the particle rifle."

It took her a second to catch his point but she finally answered, "Oh, *that*. Nah, that was on me the whole time."

"How do you figure?"

"That was an unlawful order I had you give me. And I knew it." Decker laughed as she explained, "I didn't actually have to obey it."

"Then why did you?"

"I got sick of you implying that I was withholding the rifle because I was on some petty power-trip." She tucked some stray strands of hair into her crown of braids, the unusually messy hairstyle betraying the wear-and-tear of her long day. "Look, I can be a vindictive bitch – there's no denying that – but I don't screw around when I'm leading a mission. No matter what people say, I earned both my stripes – the hard way."

"I wasn't suggesting otherwise. I just wanted...." At this point, Tal couldn't actually remember why the rifle had seemed so important to him. Feeling slightly foolish, he said, "I'm sorry if that's what you thought I was doing. It's clear you're a valuable officer. I don't think anyone disputes that."

"Oh, really?" She gave a humorless snort of laughter, asking, "You haven't heard the Corps joke about my stripes?" Decker pointed to the two silver bands on her uniform wrist, identifying her as first lieutenant, saying, "This one's called Decker, and this one's called Ricci."

"That's not fair at all. If anything, your fath—" He caught himself at her immediate frown and amended, "If anything, *the captain* can be a little hard on you at times."

"Doesn't seem to be the general consensus."

His conscience nagged at him as maybe being lumped in with that crowd, Jacoway offered, "You know...what I said about you getting by on your surname? And about your mother? That was all *way* outta line." He rubbed the back of his neck, admitting, "I've got a temper. It gets the better of me sometimes." Jacoway didn't add that Decker alone could aggravate him to that remarkable extent.

"Yeah? Me too. And I guess I provoked that temper of yours." She tossed her head, saying, "You know, seeing as how I hijacked your simulator and...." Some genuine humor warmed her voice as Decker finished, "...tried *everything* in my power just to beat your record."

"*Tried* being the operative word. I saw your data – you failed pretty spectacularly."

"Sure did," she agreed tiredly.

"Though who knows...." Tal said, tapping his neck where the computer port was situated. "Maybe if you had one of these."

"Maybe. But I need an extra hole in my head, like an extra hole in my head." A small but sincere smile lit Naiche's bronze complexion as she said, "You know, Jacoway, we keep this up and we're going to throw away *years* of mutual animosity, resentment, and scorn."

"I think...I'm okay with that," Tal said. He extended his hand. "Friends?"

Decker took it and shook. "Friends."

"As your new friend I'd like to say—"

"I know, get some sleep."

"Yeah that, but also...." Jacoway hesitated but forced it out. "What you said about having deaths on your conscience?"

Decker got very still but only answered, "Yeah?"

"The war took its toll on all of us. But I know you...had it especially rough." When she just gave the smallest of shrugs, Tal advised, "I hope you availed yourself of the counseling services available at The Rock."

"I appreciate you saying that, but *N'daa* medicine can't heal a Chiricahua mind." Before he could marshal his strenuous objections to that statement, Naiche added, "I avail myself of healers when I go home."

"We're all just *human*, Decker."

“Yes, we are. But when Doctor Clemente needed some cells to help me through an injury, she was very careful to get ones that matched my body. I think my mind deserves the same consideration.”

Jacoway was sure Decker was wrong but he was unable to articulate a valid response so he just wished her some sound sleep and headed to his own quarters.

Chapter 7

A World of Illusion

"We live in a fantasy world, a world of illusion. The great task in life is to find reality." Iris Murdoch

Ricci heard the chime at his door and called, "Enter," knowing it was most probably Zache. She walked into his quarters and hesitated, obviously having a look around. "Well?"

"Not too shabby, Old Man." She surveyed the sitting area, adding, "You have two full rooms?"

"Three, if you count the bathroom."

"Wow, a private bathroom. No wonder you always wanted the gold braid. Do you have the only one on *Lovelace*?"

Matt answered with a wry chuckle, "No, it's me, Lindstrom, *and* Kayatennae." He gestured towards the small table, which had been set up in the sitting room. "Food's not here yet, but I can offer you a drink."

"If you have a wet bar in here, I just may have to smack you."

"Well, fortunately for me, it's only three lonely bottles."

"I'll wait for dinner. We *are* having wine?"

"Of course."

"Does everyone onboard get to have a private dinner in the captain's quarters?"

"All senior officers do, at least once. It's usually the way I welcome new recruits."

"No junior officers?"

"Only one." Matt appreciated that Zache didn't feel the need to prod further, perceptive enough to guess that the one had to have been Naiche, after their reconciliation. The door chimed again and it was the kitchen staff with their dinner.

While they set up, Zache busied herself inspecting the mementos he had on a shelf. Once the staff had left, she pointed at a holo-frame and said, "Who *are* these infants?"

Matt went over to check it out and laughed as it was an image of him with Zache and Nik at their wedding. "How did they ever let you two get married that young?"

"You really want your mind blown? We were the same age the twins are now."

"Ouch, now I *do* feel old."

Zache turned her attention to the next holo and whispered, "Well, would you look at that." It was an image of a young Matt with his arms around Naomi Decker.

"Yeah, one more advantage of everybody knowing the truth is that I could bring things like that out of hiding."

She picked up the frame for closer inspection, commenting, "My God, Naiche looks like her mother."

"True, but the Fwalayna Noctay did pick up on the fact that we were related." Matt couldn't repress a small smile at the memory.

"They thought you two looked alike?!"

"Smelled alike," Ricci corrected. With a sigh, he added, "Naiche said, 'Hey, only when we play basketball.' *That's* when I told her she could take a break."

Zache laughed and replaced the holo-frame on the shelf. “When was that taken?”

“On a camping trip we took right before the war started. We snuck away to Centauria for a whole week.”

“Nik always said you two were more than friends. I, on the other hand, said if that were the case – you’d tell us.” She raised an eyebrow at him. “Having no reason to hide it from us.”

“Our reasons were complicated.”

“You *made it* complicated.” Matt braced for a renewal of Zache’s litany of complaints about him having hidden that relationship from her for decades; therefore, he was wholly unprepared when she abruptly changed the subject, asking, “Do you have any holos of you and Jackie?”

“Smooth, Zache, real smooth.” He moved over to the table, saying, “Shall we eat?”

Zache joined him at the table but said, “I’m serious. I always liked you and Jackie together.”

While pouring generous glasses of wine for both of them, Matt stated, “We were never *actually* in love.”

“So you always said.”

With a sad shrug, he explained, “I was still mourning Naomi; I believe she was still mourning her husband.” He paused briefly as memories came rushing at him. “But it *was* good, while it lasted.”

“If that’s true – then what went wrong?”

As Matt took the covers off their plates, he struggled to remember the details. “Our jobs got in the way, mainly. The war turned even uglier; we were both chasing the captaincy...all the usual Corps shit that derails relationships.”

“This *is* affecting you, isn’t it?”

“You mean, did my history with Jackie affect my decision to go to Antares?”

“I meant what I said.”

"Of course, it's affecting me! I lost enough friends during the war. I can't stand the thought of losing another this way." He looked Charani in the eye, assuring her, "But, that isn't why I made my decision. Every instinct I have, from all my years of command, is telling me that going to the Antares system is the right move."

Zache seemed unconvinced but lifted her wine glass in a salute. "Well, gold braid decides."

"Yes, and I decided that we need to know what happened with the *Daleko* before you and Jacoway go in." He leaned forward and said, "I plan to do everything in my power to ensure that *Jackie* is the only old friend in danger from this mission."

Though she seemed touched at his concern, Charani nevertheless shook her head, musing, "Knowledge like that could be invaluable, but it won't do us a bit of good if it comes too late to reach the Expanse."

"Well, then, gold braid gets the blame, too. That's the way it works."

***

Lovelace had reached the small, attractive planet, Scorpii-d, which matched the coordinates from the distress signal emitter. The bridge staff was eagerly awaiting word from Lateef as she scanned for signs of Bastié, Pilecki, or their ship. "There's nothing, Captain. I'm finding technology and alien life-signs – many different ones – but I'm reading nothing of Terran origin whatsoever."

Decker shared a look of disappointment with Kennedy while Ricci asked, "Weaponry?"

"Nothing at all like that."

The captain turned to his linguist, "Any progress on decoding that transmission we've been receiving, Sasaki?"

"Yes, sir, I've translated the entire thing."

Ricci's attitude went from frustration to admiration in an instant. "That's remarkable progress, Commander. Congratulations on the stellar work."

"I wish I could accept the accolades, Captain, but this communiqué was obviously designed to be very easy for any advanced species to translate. It's a welcome message...or invitation, maybe...to all comers."

"Invitation," Decker said. "What is Scorpii-d, some kind of round-the-clock party planet?"

Sasaki couldn't hide a small smile at her question but answered, "More like a way-station, I think."

Lindstrom asked, "Do you mean to say we've stumbled across an inter-stellar rest stop?"

"Pretty much, Commander."

"How 'bout that," Ricci said. "So, the *Daleko* could have been here, at one time?"

"Makes sense," Charani answered. "They could have stopped here, activated the mobile emitter because there was some problem but then fixed their ship and gone on their way."

"That doesn't add up," Kennedy objected. "How badly damaged would the ship have had to've been for them to bother with the mobile emitter? And then they just fixed it?"

Decker continued, "And where are they now? If they fixed the ship, they should either be in the vicinity or have made it to a relay station and gotten a message to Uniterrae by now."

"Maybe we just missed them," Lindstrom said. "We haven't been near a relay station, ourselves, in weeks. We could *literally* be two ships that passed in the night."

"Could be," Ricci answered. "Strange that there's no trace of them left behind, not even a fuel signature. How good are your scans, Lateef?"

"They're crystal clear, Captain. I rarely get scans of this quality. There's no interference from the planet's base technology, or any of the aliens present, or even any stray radiation from Antares."

Kennedy suggested, "Maybe Decker and I should take a shuttle down. Do a quick sweep over the planet's surface. Whata you say, Captain?"

"How much time are we going waste here before we accept that we just missed them?" Jacoway asked.

"I understand you're anxious to get to the Expanse, Commander, but I think we can spare a little time to do a thorough search," Ricci said. He glanced over at Zache adding, "Since we made the trip and all." He then turned towards the Tactical officers. "Kennedy, take a small reconnaissance team down. You've got an hour. Use it wisely."

"Yes, sir. Ready, Deck?"

"You bet! There's gotta be more to this planet than what those scans are telling us."

"Why do you say that?" Lateef asked.

Decker paused on her way towards the exit, explaining, "Because of what that Darden thing called it. You know...when Commander Sasaki finally translated that stream of screeches." She turned to Sasaki, asking, "Didn't you say that it called the place where it found the emitter the 'bountiful hunting ground'?"

"That's right," Jeff answered. He gestured at the blue and green planet visible on their view screen. "That's not bountiful enough for you? Ships must land there all the time."

"Seems like bountiful hunting for a predator. But for a scavenger? Not so much."

Jacoway asked, "What makes you think the Dardanze is a scavenger?"

"Well, just look at it for one thing. It had no weapons or even offensive capabilities. And its only defensive maneuver was running away and hiding. That's not the way a predator acts."

Lindstrom said, "We don't know what it meant, though. What would a 'bountiful hunting ground' even look like for something like the Dardanze?"

With a shrug, Decker replied, "I don't know...something that would be the equivalent of a desert in space. Some hellhole full of wrecked ships. That's what I was afraid we were gonna find here."

She turned to go but ran right into Con's outstretched arm. "Hold up. What did you just say?"

"I said that from the Darden's description I was expecting to find a much more dangerous planet."

To her confusion, Con didn't reply or move. After a second, he looked over at the science console and asked, "Aqila, how good did you say your scans are?"

"I've never seen better, why?"

"Never?" Ricci asked.

"No. They're like the ones they give you in first year Telemetry to learn on."

Ricci was nodding his head while studying the planet on the view screen. "Could something like that be faked?"

Charani objected, "Oh, come on, that's ridiculous—"

Moving back towards the view screen, Kennedy interjected, "Good way to set a trap, wouldn't it be?"

"You kind of see traps everywhere, don't you, Commander?" Jacoway asked. "You thought the initial distress signal was a trap, the Dardanze's ship was a trap—"

"That's the way a chief tactical officer thinks. Paranoia is the main part of my job," Kennedy laughed.

While they were talking, Lateef had been busily re-checking her scans and sensor readings. She looked up, saying, "These scans *are* odd! There's no ebb and flow of signal. There's no change in resolution at any point. If I plotted it out – it'd be a straight line." Aqila brought up a holographic screen with the data displayed. "See?"

Lindstrom asked, "What does that mean, exactly?"

Though seeming reluctant to propose the idea, Aqila eventually explained, "Well...what we're receiving as sensor results...it *could* be a projection of some sort."

While a murmur of concern and surprise went up from the bridge crew, Ricci ordered, "See if you can either confirm or deny that supposition, Lateef."

They waited in tense silence while Lateef worked at her console. Finally, she brought up a screen showing more sensor data, saying, "No matter how I calibrate, tune, or tweak the sensors, these are the *only* results I'm getting." She looked over at the captain. "If this is a projection – it's an impenetrable one. I'm just not sure."

"Okay, then. Forewarned is forearmed, at least." Con glanced at Decker and they nodded in unison while Con finished, "We approach Scorpii-d with extreme caution."

"No, you don't." Ricci announced.

The tactical leads looked at him in surprise. Decker asked, "We don't approach with extreme caution?"

"You don't approach *at all.* Not until we know for sure what's waiting down there for us."

Zache asked, "How long will that take?"

"I wish I knew – but we need a definitive answer before I'm sending a team down." Ricci ran a hand through his hair and turned to Lateef. "What can you do if I give you four hours, Commander?"

"My best," Aqila answered.

"Put every available science officer on it," Ricci said. "Let's ensure your best yields an answer."

"Aye, sir."

***

Three hours later Decker was having dinner in the mess hall when Ricci appeared at her table, tray in hand. "Eating alone? Where's Kennedy?"

"You know, we're not actually joined at the hip."

"No, of course not." Ricci sat down, asking, "So, where is he?"

"He brought Aqila some dinner. He'll be here in a few." She smiled at her father saying, "I could turn the tables on you and ask where Zache is."

"She's holed up with Jacoway."

"She's pissed about the delay, isn't she?"

"Commander Charani is concerned that we will miss our opportunity regarding the Expanse." Ricci took a bite before admitting, "So yes, she's pissed."

"What is with those two? They can't wait to charge head-first into that death trap."

"You mean there's actually some danger *you're* not looking to dive into?" Ricci sipped his drink, muttering, "Hallelujah."

"We're not *sure* Scorpii-d is dangerous. Con and I just want to—"

"I know what you want, this isn't up for discussion."

Naiche ate in silence for few moments before asking, "What happens in an hour if Aqila doesn't have a definitive answer for us?"

"I will render an appropriate decision when we come to that juncture."

"Huh, so being a captain means just making it up as you go along?"

"You've cracked the code," he answered drily, as Kennedy joined them at the table.

"What did Aqila have to say?" Naiche asked Con.

"And I quote: 'Thanks for the sandwich, now go away, I'm busy'." Kennedy nodded towards Ricci. "Did you plead our case for not waiting?"

Matt shook his head in exasperation. "You two are as bad as Charani and Jacoway."

"Now, sir, I'm not sure that's quite fair."

"Yeah, we're much worse than them," Decker said.

"Deck, stop helping."

"The captain is not going to change his mind."

"No, I'm not," Ricci confirmed. "As I just told Lieutenant Decker, my orders are not subject to negotiation. Now, can we talk about something else during dinner?"

"Sure." There was an awkward silence before Kennedy ventured, "Umm, so, what did you guys get Sasaki and Avery for their wedding?"

Decker looked up in alarm. "Shit, when is that?"

"Well, that answers that question. It's less than three weeks after we're scheduled to get back."

"Whoops. What'd you and Aqila get them?"

"We ordered a set of ceramic knives from *Nihon* that Sasaki wanted. How 'bout you, Captain?"

"I was planning on going shopping when we got back."

Studying Ricci though narrowed eyes, Decker asked, "Were you? Or were you planning on sending *someone else* shopping?"

"How about I give *someone else* the money to purchase a very nice gift from both of us and then she can just sign the card?"

"Deal."

Con banged his fist in mock disgust. "Buy your own gift, cheapskate! What're you gonna do with all that money you've been saving up?"

"I'm gonna treat myself to something spectacular – like my own QNS simulator so I can take all the joyrides I want." Naiche felt the pressure of Con's boot on hers and said, "Stop nudging me. He knows."

Ricci glanced between them and asked, "You told Kennedy, too?"

Con jumped in before Decker could answer, assuring him, "Yes, sir. I, uh, had a talk with her about it."

"Uh-huh."

At that moment, the nearest VICI unit buzzed and announced, "Captain Ricci, your presence is required in Science Laboratory-One. Priority situation." Ricci acknowledged the announcement. As he got up, he said, "Looks like we may have an answer," and left.

When he was gone, Con asked Naiche, "You think he bought that?"

"About as much as he planned to buy that wedding gift himself."

Ten minutes later VICI issued similar directives to Decker and Kennedy. They walked into the lab in time to hear Ricci say, "Decker was right. Though hellhole may be underselling it."

Chapter 8

Hell and Destruction

"Persevere even though Hell and destruction should yawn beneath your feet." Percy Bysshe Shelley

Decker watched Lateef enlarge the schematic visible on the lab's holographic screen. "As I explained to Captain Ricci and Commander Lindstrom, this is a pretty crude model, it probably looks like a chalk drawing to you, but at least it's accurate. It provides a clear indication of what we're facing on the planet, Scorpii-d. The real planet – not the projection."

Charani asked, "How were you able to penetrate the projected data?"

"We ended up removing all but the neutrinos from the lepton particles of the sensor arrays. The neutrinos can penetrate but unfortunately the data we get using only neutrinos aren't nearly as detailed."

After studying the black and white screen for a minute, Decker asked, "That can't be liquid in those canyon-beds, right?" She pointed at the area where dozens of ships were visible in the canyon bottom.

"Actually, it scans as a liquid – of sorts."

Con said, "If it's anything like water it must be really shallow, then."

"No, it appears to be fairly deep. The liquid is highly dense and seems to be extremely viscous."

"Both dense and extremely viscous? That's a contradictory state for a liquid," Charani objected.

"Yet, it apparently exists," Lateef said. "It can support the weight of those ships but I'm guessing that anything that lands on it gets stuck and maybe even is eventually drawn under."

"Is one of the ships the *Daleko*?" Kennedy asked.

"Yes," Aqila answered, while drawing a circle around the ship in question. "This one matches its specifications."

"Shit," Decker said, pointing to the ships and wreckage surrounding the *Daleko*. "It *is* a ship graveyard." She studied the low canyons walls, which were steep and spiraled up into sharp peaks. "What's the terrain like?"

"Dry and extremely friable. There's not much vegetation at all."

"Can you read life-signs?" Jacoway asked.

"Yes. We're reading tens of thousands of life-signs all over the planet, several hundred in the 300-kilometer canyon that contains the *Daleko*. But the reconfigured sensors aren't capable of distinguishing human from alien."

"We won't know if Bastié and Pilecki are still alive until we get down there and check it out," Ricci said. "Question being, how do we get a shuttle anywhere near the *Daleko* without getting stuck in that sludge ourselves?"

"We'll have to land on the surrounding area," Decker said.

"Which parts? That narrow beach area," asked Lindstrom, pointing out a strip of land no more than two meters wide, "or perhaps the cliff-face, or those jagged tops?"

"Landing the shuttle is only problem number one," Kennedy said. "The bigger question is – who or what set up this spider-web of a planet and why? What are they doing with their victims?"

"I don't think anyone set it up," Lateef said. "I think someone is just taking advantage of the naturally occurring...um, trap that Scorpii-d creates."

"Still, their intentions are hardly benevolent," Decker said. "They're definitely luring ships down there for a reason. And it ain't to find tourists for their gift shop."

"Right. So, we want to engage them as little as possible. Which means moving fast." Kennedy studied the model intently before saying, "Our best tactic to get in and out quickly would be to land the shuttle right above where the *Daleko* is. Then we can either rappel or climb down the canyon walls depending on what we find." He moved his head closer, musing, "It looks like there's a lot of ledges, so there might be a path—"

"Right above the *Daleko* where, Commander?" Lindstrom asked. "On those peaks? How can you do that?"

"Three points define a plane, sir," Con answered with a smile.

"What the hell kind of pilot could do that?" Zache asked.

Con nodded at Decker, saying, "The best."

"On this trip, that would be me," Jacoway asserted.

"Not for this mission," Kennedy said. "Deck's the one for this job. This is like during the war—"

"What? Do you Tactical people think fighter pilots weren't part of the war? Do you know how many times I took fire? How many times our squad—"

Ricci interjected, "No one's questioning your skill or valor, Jacoway. You're just not the appropriate pilot this time around. Kennedy's right; Decker's needed here."

“Oh, I see how it is,” Tal said, as he folded his arms in dejected defiance.

“Do you?” Ricci snapped. “Well, I’m not in the habit of explaining my orders to *anyone* but I will do so....” The next three words were bitten off, creating a verbal stop in between each one. “...just this once.” He faced Tal directly asking, “How many times did you land your plane *anywhere other than the hangar*?”

Jacoway clearly got the implication immediately and answered, in an abashed tone, “Umm, never, sir.”

“Right. And do you know what Micro-craft pilots did every *goddamned* day?”

“Landed their planes in a war zone, sir.”

“Correct, again, Commander.”

Tal drew himself up to his full height, and offered contritely, “My apologies, Captain, my emotions got the better of me there.”

“Your emotions? I think it was your ego,” Ricci declared. In a slightly easier tone he added, “But either way, apology accepted.” He turned to Kennedy, saying, “Assemble your team, Commander.”

“Yes, sir.” Con asked Aqila, “How long will it take to adjust the shuttle’s sensors to neutrino only?”

“No more than a few minutes. I’ll put Kaplita on it immediately.”

“Great. Let’s go, Deck.”

Decker followed Kennedy at a brisk pace towards the exit; both were delayed when Ricci called out, “Kennedy!” When they paused and faced the captain, he said, “This is in no way intended to be a suicide mission. If at any point things look dicey, you pull the plug. Am I clear?”

“Crystal, sir. Let’s go, Deck. You see to the shuttle; I’ll round up Bayer and Abello.”

***

"Almost got it," Decker said as she stared intently at the shuttle's data screen, in her fourth attempt at landing on the canyon peaks.

Kennedy was watching her progress from the co-pilot's seat and asked, "Is this trickier than that time on Toliman when you landed your Micro-craft in that mountain crevice?"

With a dismissive laugh, she answered, "Waaay trickier. And this crude sensor display sure isn't helping any."

"At least you're not under fire this time, right?"

"You got a point there. Let's just call it, a different kind of tricky – like trying to balance a dinner plate on some sticks...." While carefully working the controls, Deck muttered, "Except nowhere *near* that easy."

Kennedy held his breath as she delicately maneuvered the shuttle into place, attempting to find a level, steady position. But yet again, when the shuttle landed it immediately started listing, this time to the port side.

"Damn it!" Decker snapped and moved the shuttle back up into the air.

Kennedy started wondering how many tries he could possibly allow before aborting; she'd been trying for almost an hour and her braids were damp with sweat. Moreover, they were sure to attract some unwanted attention from the locals soon. The problem was, if Decker couldn't land the shuttle safely on Scorpii-d, he didn't believe there was a soul on *Lovelace* who could – not even Jacoway. As Decker declared, "Fifth try's the charm," Con privately decided, it was the last.

Twelve apprehensive minutes later, he felt the shuttle come to rest again and waited for the inevitable slide. Everyone in the shuttle, including Kay, seemed frozen in place, awaiting certain disappointment. However, the shuttle didn't shift position, even when Con got up and jumped around.

Decker threw her arms up in a victory salute and yelled, "Yes!" She started studying the view screen and sensor data. "I'm not detecting any life-signs in the vicinity."

"Aqila said we might not be able to trust our sensor readings regarding *all* life-signs."

"I know – proceed with caution."

Con leaned over her shoulder, analyzing the output on the screen. "Looks like we'll have to rappel out of the shuttle about ten meters to the closest ledge. Then we should be able to follow that narrow path down to the *Daleko.*" He stood up and stretched, saying, "Before we go, Deck, see if you can contact *Lovelace*¸ let 'em know we were able to land safely."

"Will do."

When Decker reported nothing but static, Con simply sighed, announcing, "Okay, we're on our own." It was not an unexpected report; Aqila had also warned that transmissions might not penetrate the data projections from this planet. Kennedy strapped on his daypack, ordering, "Let's get to it; daylight's burning. Abello, you stay with the shuttle, I don't want to leave it unguarded. Bayer, you're with me and Decker."

While tossing a bulky pack to him, Decker said, "Con, hang onto the rescue kit for me until I get onto the ledge." She quickly secured and dropped a line out the back hatch. After hooking Kayatennae's rappelling harness into her own, she lowered both of them out of the shuttle and onto the ledge. The other two soon followed, Bayer carrying all three particle rifles.

They redistributed their gear and faced the descent ahead. The rock canyon was of a crumbly, dull-white stone and the most direct path they found was harrowing, requiring careful navigation and frequent use of camming anchors but after twenty minutes, they made it down to the narrow beach.

Among the wrecks and intact ships, the trio quickly located the *Daleko,* Con immediately observing, "The hatch is open."

Decker pointed to the ship next to the *Daleko,* resembling a giant black and orange stingray. "That ship is moving. Might be our alien hosts – boarding ships."

Con watched for a few seconds, before responding, "Could be. Looks like they're trying to get that ship out of there." The team waited for a tense ten minutes, weapons drawn at the vessel in question, but there was no approach from its inhabitants who seemed solely intent on freeing the ship from its adhesive landing pad.

Finally, Kennedy decided to take the risk of inspecting the *Daleko*. He used his hand-scanner to check the ship but found nothing. "No life-signs – for what it's worth." He looked at his team. "Bayer, you stand guard. Let us know immediately if you see anything come out of that other ship. Deck, let's go in. Weapons up."

Each of them, including Kay, took a short running start to cleanly jump into the open hatch, neatly avoiding the surrounding liquid, which shone and glinted like a sea of ebony. It didn't take long to confirm that the *Daleko* was deserted though it appeared to have been mildly ransacked. "Wonder if Bastié and Pilecki—"

"What the hell is this?" Decker exclaimed. "Kay, stay away from it!" she ordered.

Con turned to see what she was talking about. "It's a big pink ball of...something?"

After poking at it gingerly with her rife, Decker said, "Dust. It's a meter-high ball of pink dust." She looked at Con saying, "What the fuck is a giant ball of pink dust doing on this ship?"

"Those transmissions from the *Daleko* when they aborted the mission! They said something about a dust storm. Remember?"

"How the hell would it have gotten inside? And why is it all in a ball?"

"I don't know; it's not our biggest concern right now, either. See if you can get this ship to move," Kennedy said, suddenly grateful for her rule-flaunting ways.

She hopped in the pilot's seat, donning the transdermal band that she found after rooting around on the floor for a few moments. After several dozen tries, she turned to him, reporting, "No way. The propulsion system is working fine but I can't get the slightest bit of lift; it can't overcome the drag from this...black muck. This ship is not going *anywhere*."

Con was disappointed but not surprised. "Okay. It's more important to see if we can locate Bastié and Pilecki, anyway." He pointed to the dog. "Let Kay sniff the cockpit and then we'll see if he can find any traces of them outside."

Once they exited the *Daleko*, Bayer confirmed that there had been no change in activity from the neighboring ship so they concentrated on the search for Bastié and Pilecki. Kayatennae soon caught traces of his quarry and the human trio started following him. Abruptly, Kay stopped and scented the air, his ears pricked, and his body tense. He stared off in the distance, emitting a low growl.

"Someone's coming!" Decker warned.

"You think it's Bastié or Pilecki?" Bayer asked.

"Definitely not," Deck said. "Whatever it is, Kay doesn't like it. We should take cover."

They found shelter behind a jagged piece of the canyon wall. Kennedy motioned the others to stay under cover but peeked over to keep watch. After a short wait, three aliens came into sight. In the growing dusk it was hard to make them out, but to Con's eyes, they looked like giant crabs formed out of quartz. What concerned him most was the major weapons they were carrying. The guns were large and the aliens, who were headed for the ship sitting beside the *Daleko,* had them at the ready. To his amazement, the aliens easily skated across the black sea and started boarding the strange looking craft.

Con ducked back down and whispered to Decker, "You're not gonna believe this—" He was interrupted by a volley of photon bolts, letting him know that the aliens had excellent hearing.

"What're we looking at, out there?" Decker asked.

"Um, three giant crystal crabs, walking on two legs, armed with, apparently, some kind of photon grenade launchers."

"Please tell me you're making at least some of that up."

The team readied their weapons, prepared for a full out assault, but heard only the continued sounds of the alien ship's power drive. When Con looked out again, the aliens and the ship were gone. He stood up to check and confirmed his first impression. "What the ever-loving fuck?"

Deck stood up, weapon drawn, but lowered it slowly saying, "Where did they go?" She pointed at the spot where the alien vessel had been. "And how *in hell* did they get that ship out of there?"

"Maybe that vessel could overcome the drag?" Bayer asked.

"Maybe," Con said as he scanned the darkening sky. "But how did it disappear that quickly?"

"I don't know – but I'm betting the reason isn't good for us." Decker said. "Come on, Kay, let's find Bastié and Pilecki, and get the hell out of here."

However, after a few steps onto beach area, they were close enough to make out the sight of five alien corpses. Bayer turned the beam of her wrist light onto the scene of mass death. The bodies looked like they'd been formed of thick brown rubber tubes, now leaking thin white fluid.

"Oh," Bayer said. "They weren't firing at us."

"No. Not yet, anyway," Con answered. "Let's see what we can do to keep it that way."

Pointing to the alien bodies, Bayer asked, "What if they're not dead yet? Shouldn't we try to help them?"

Deck had already fallen to her knees and was examining the alien bodies. "I'd like to – but how? We know nothing about their anatomy...nothing about their physiology...nothing about them at all. I could end up doing more harm than good." She must have realized it would be difficult to truly harm them further, since she amended, "Well, all I could probably do is hasten their deaths. I mean, if they're even still alive; I'm not even sure of that."

"You're right, so let's get moving," Kennedy ordered. "Let's help those we can – Bastié and Pilecki."

Decker nodded sadly, and jumped to her feet, giving Kay the command, "Find people!" The dog once again was on the hunt for the missing humans. He quickly regained the trail and eventually led them up a short, gradual path to a small opening in the canyon wall. Kennedy ducked down and flashed his wrist light into the darkness. Towards the back wall, he thought he saw two indistinct mounds. When he crawled forward, his light shone on the unmistakable navy-blue of UDC uniforms. "Captain Bastié? Commander Pilecki?"

One figure propped himself up on one elbow. A pale, unshaven face blinked uncertainly at Kennedy for an instant, finally croaking out, "I really hope this isn't a dream."

"Commander Pilecki? I'm Lieutenant Commander Kennedy. I'm here with a rescue team from the *Lovelace*."

He simply whispered, "Thank God," before slumping back down onto the ground. Con was unable to arouse either figure again. He called out, “Decker, get your med kit in here, pronto! These guys are in pretty bad shape.” He didn’t feel it necessary to add that he was unsure if Bastié was even still alive.

Chapter 9

Like Land and Sea

"Love and anger are like land and sea: They meet at many different places." Patricia A. McKillip, The Changeling Sea

Half an hour later, Pilecki was sitting up, sipping from a water tube, and munching on a meal bar. Deck had said he was mainly suffering from dehydration, exhaustion, and mild malnutrition. Bastié was in much worse shape, and Deck was still at work on her.

"Are you up to telling us what happened here?" Kennedy asked

"I think so," Pilecki answered. "The short version anyway." He looked towards the exit of the cave where Bayer was standing guard, asking, "Are we going to be able to get out of here? Where did you guys land?"

"We're up on the canyon, nowhere near that black muck. We'll be able to get out. As soon as Decker says it's safe for us to move Bastié." He glanced back at Deck who, after studying the scanner in her hand, applied some kind of med-patch to Bastié's forehead. "What happened to her?"

"That short climb up here just about finished her. She was in pretty bad shape after the Expanse." Pilecki shook his head. "We badly underestimated the QNS's neural strain on a pilot in the quantum entanglement. Twenty minutes in and Jackie was already showing massive fatigue and cognitive impairment. You could almost smell smoke coming from her neural port."

"That's when you decided to abort?"

"It sure as hell was. Somehow Jackie managed to get us out of there. It damn near killed her but she managed it. Even after we ran into that dust cloud."

"The pink dust?"

"Yeah, how did you know?"

"We were on the *Daleko* – there's a ball of that stuff there. You picked it up in the Expanse?"

"Yeah."

"How did it get inside?"

"No fucking clue. One minute it was outside the ship, then the next it was all over everything. Especially us."

"Is that what you meant when you reported being under attack?"

"Yeah – our initial concern was that it was some kind of weapon that had been launched at us from...somewhere. It soon became clear that it was pretty innocuous though we were baffled how it had possibly breached the hull. At that point, our major concern was getting out of the Expanse; we figured we'd deal with it later. When Jackie and I were leaving the ship, all of it slowly contracted into a ball. That stuff definitely has some weird properties."

"You were able to get off of the *Daleko* after the aliens boarded?"

"No, before that. I might not be able to understand their language but their weapons told me loud and clear what their intentions were. When I saw them heading for the *Daleko,* I figured we had no choice but to abandon ship. I wasn't going to be able to hold them off alone and I had Jackie to worry about, too, so I grabbed the bail-out bag and we left before they got to us."

"Good choice. We saw first-hand what happens to those who refuse or try to fight back."

"I figured. We saw some corpses on our way here, too. Also, some other stranded travelers. Based on the shape they were in, the only choice this shithole planet presents is instant death or slow death. Those crabs didn't even bother chasing after us; they only care about getting the ships – the passengers can die fast or slow, they don't seem to have a preference there."

"Do you know how they're getting the ships out?"

"No, I haven't been able to watch them do it. Whatever their method is, it's quick."

"What're they doing with all the ships? Any idea?"

"No." He shrugged, venturing, "Selling them, building an armada...I don't know." With a small bitter laugh, Pilecki added, "It's probably royally pissing them off that they can't figure out how to fly the *Daleko*. I was afraid they might come after us for help with that. That's why I've kept us holed up here."

"Why did Captain Bastié choose to land here?"

"It was me, I was piloting by then, using the transdermal link. This place presents very differently from orbit."

"Yeah, we know. We figured out that what you see from space is all a projection."

"Good for you. I didn't. To me it seemed like a good place to rest, re-group. Umm, clear my head."

"Clear your head?"

"Yes, I guess the strain was getting to me, too. We came out of the Expanse *nowhere* near where we'd been. My plan was to get us to a relay station to send a message but...frankly, I just wasn't thinking clearly. The ship was...you always have feedback from the ship with the QNS but after the Expanse...it got weird. There's no other way to describe it."

Con put a comforting hand on the agitated man's arm. "No need to. It's clear you've been through hell. How long have you been here?"

He slowly pulled his hand-held out of his pocket and consulted it. "Nearly eleven days."

"Before we showed up, how long since you'd had food or water?"

Pilecki ran a hand over his unruly blond hair, obviously considering the question. "The bail-out bag has a week's worth of supplies but I stretched them out as best I could. We ran out completely about...two days ago. I was beginning to wonder if it mattered. I pretty much gave up hope of rescue when the mobile emitter disappeared. How did you find us?"

"That's a long story – I'll tell you on the shuttle." Kennedy looked at Decker. "Deck, can we move Captain Bastié soon?"

"Yeah, I'm going to have to carry her but we can move her in a couple of minutes."

Pilecki looked from Kay to Decker and asked, "You're S and R?"

"I was – for a year," Decker answered. "During the war."

"Deck?" Pilecki asked, confirming what he'd just heard Con call her. "Holy shit! You're Naomi Decker's kid, aren't you?"

Kennedy longed to say, 'No, she's the woman who helped you and is probably saving Bastié's life right now,' but couldn't speak so to an injured man. Anyway, Decker had already answered in the affirmative.

Con's ire doubled when Pilecki asked him, "You said the *Lovelace*, right?" He didn't wait for an answer but just said, while looking thoughtfully at Decker, "That's Ricci's ship, isn't it?" Con reminded himself that people didn't really know what it was like for Deck, having her parentage brought up constantly. But Kennedy knew, and it bothered the hell out of him.

Rather than responding to Pilecki's query, Decker said to Con, "Help me carry Bastié outside where I can get her into the rescue harness."

Once she had readied Bastié and they were all out onto the ledge, Decker took out her oculiscope and tuned it to night vision. She surveyed the steep climb up to the shuttle and then looked down at Kay. "Looks like there's portions of this trip, I'm gonna have to carry both of you. Well, nothing I haven't done before."

Con asked Pilecki, "You think you're up to the climb?"

He looked doubtful, but said resolutely, "I guess we'll find out. Wish it wasn't so dark."

"Deck, can Bastié's transfer possibly wait until it's light—"

Bluntly, Decker broke in, "No, I don't think we should. I have her stabilized but I'm not exactly sure what kind of neural damage she's suffered – or how much. She needs immediate medical attention. The kind I can't give her."

Pilecki declared, "I guess it's best we go in the dark anyway. If those crab-aliens spot the shuttle, they'll want it. Badly. And they can climb like monkeys." He zipped up his uniform with a shaky hand, asking, "Do you have extra cam anchors?" When Con handed a pair to Pilecki, he said, "Thanks, I've done some rock climbing. I should be okay."

"Good, let's get going," Con said. He briefly considered carrying Bastié himself but she was a slight woman and he figured he'd be carrying or supporting the much heftier Pilecki some of the way.

They made an agonizingly slow climb up with Bayer in front, lighting the way, and clearing the path as much as possible. Every time they stopped to let Pilecki rest or to check on Bastié, Kennedy used his oculiscope to check for near-by crab-aliens. He never spotted any nor did sensor readings pick up anything but he couldn't rid himself of the nagging feeling that they were lurking just out of sight.

The team was navigating a particularly steep area when Pilecki's cam anchor came loose. He lost his footing and started sliding down the rock face, heading for the canyon bottom. Kennedy swiftly repositioned himself so that he was able to grab Pilecki and stop his descent, saving him from near-certain death. Con gripped his own anchor with all his might, realizing, at that point, it was the only thing standing between them and a disastrous fall. He fumbled around and managed to sink another anchor in with his free hand, which gave them some additional stability, but also hampered his doing more than continuing to grip the rock face.

Kennedy wasn't sure if Bayer and Decker, above them by a good ten meters, were even aware of their predicament. He weighed the danger in shouting for their help against drawing unwanted alien attention. It didn't seem wise; there had to another way. He started gingerly trying to get a hand free to activate his comm link and contact Deck.

While Con was trying to get help in an unobtrusive manner, Pilecki was desperately attempting to re-sink his own anchor. "Son of a bitch, I can't get it to stick!"

Kennedy found there was no way for him to get a free hand without risking a fall. Maybe they should just go with it. He was about to suggest executing a short, controlled fall to the ledge below them – a desperate act Con wasn't sure Pilecki would survive unscathed – when he heard a rustling sound nearby. Kennedy froze, trying to make out if there was a crab alien approaching or not.

He expelled an immense sigh of relief when Decker made a sudden and unexpected appearance at his side. She had rappelled down to them, and now offered a hand, saying, “Hey, Boss, looks like you could use some help there.”

“What was your first clue?”

Deck nimbly transferred Pilecki to her grasp, secured him to her harness and then hauled them both up to the ledge where Bayer was waiting with Bastié and Kayatennae. Con joined them a minute later and they all paused to catch their breath. He asked, “How did you know we were in trouble?”

“*I* didn’t,” Deck answered; she pointed at Kayatennae, explaining, “Kay did. He’s the one who alerted me that something was wrong below us.”

Kennedy stroked the dog’s head. “Thanks, Kay. You sure are a life-saver.”

“I think that designation belongs most properly to *you*, Commander,” Pilecki said. “If you hadn’t arrested my downward slide when you did...well, I think we all know what the result would have been.”

“I’ll accept that title when we’re all safely aboard *Lovelace*,” Con answered. To his extreme relief, they finally made it to the ledge just below the shuttle without further incident or encountering any hostile aliens.

Con and Naiche were both soaked through with sweat by then and they rested briefly while unzipping their uniform jackets slightly. As they were hoisting Bastié up into the shuttle Kay started growling menacingly. Before Kennedy could say anything, Decker answered his unspoken question. “I think we were followed.”

Con said, “Yeah, I’ve been thinking that all along. Those bastards are almost invisible in the dark.” They both readied their weapons.

When no imminent move was made by their potential attackers, Con climbed up the line into the shuttle and pulled Pilecki in. A photon blast exploded right next to him as he finished. He fired at the source of the shot and there was an answering volley in the direction of the shuttle.

Kennedy quickly slid back down to the ledge next to Decker, who had started firing at the aliens below them. "Looks like they're determined to either get the shuttle or destroy it!" she yelled.

"Bayer," Con ordered, "Get up into the pilot's seat. On my command, be prepared to take off without us." He waved a hand indicated that the "us" was him, Decker, and Kay.

After only a moment's hesitation, Bayer said, "Yes, sir," and quickly boarded the shuttle.

He and Deck stood back to back, as they had countless times on the field of battle. They fired into the darkness using their acousti-scopes to aim, desperate to protect the shuttle and the injured officers on it. Con yelled, "Thank God those things aren't very accurate at a distance."

"Let's not let them get any closer," Decker shouted back.

There was no more time for talk as he and Deck kept firing, expertly aiming at the source of the shots. Abello assisted, firing from the back hatch of the shuttle. They heard sounds like shattering glass whenever they scored a direct hit on one of the crab-aliens. After ten grueling minutes, the firing suddenly stopped. He waited for another uneasy, silent minute, and then Con asked, "Think we got 'em all?"

"Yeah, sure looks that way," Decker answered, between gulps of air.

Con yelled up, "Abello! How's it look up there?"

Abello flashed his light all around them. "Nothing in the immediate vicinity!"

Con breathed a sigh of relief and climbed up into the shuttle, while Decker prepared to clip Kay into her harness. Only later would Con be able to piece together exactly what happened next when Kay suddenly launched himself into the air. One of the crabs had apparently been clinging underneath the ledge and jumped up onto it, taking direct aim at Decker.

Kay hit the alien square in the chest, knocking it to the ground, and the alien's gun clattered over the edge. Decker rushed over to try and free Kay from its grasp, but before she could, the crab-alien threw Kay off the ledge into the darkness. There was sickening yelp from below and then heartbreaking silence. When Decker fell to the ground, looking over the edge, Con blasted the alien into a thousand satisfying shards.

Decker whipped out her oculiscope and focused on the area below them. She stood up, reporting, "There's a cloud of dust about a kilometer away. Could be a lot more of 'em coming."

"Yeah, those are reinforcements, I'm sure. We better get out of here." Con gestured towards the hatch he was leaning out of, firmly ordering, "Now," knowing he had to break through her fog of grief. Not seeming to have heard him, Decker pulled a rappelling line out of her pack and shoved the anchor into the ground. "What are you doing?!"

"I'm going after Kay."

"Deck, he's dead. I'm sorry but *we have got to go*."

"You don't know that. I saw him on a ledge down there. He's strong – he might still be alive!"

"I know how hard this is for you, but we *do not* have time. Kay gave his life for yours. Honor that sacrifice by getting on this shuttle – *NOW*!"

"No! I'm not gonna leave him like this."

"You saw it for yourself; there are more hostiles on the way—"

"I know, take off without me," she said as she started down the line. "Come back for us."

"Come back for you?! Who the fuck is gonna land the shuttle?" Con screamed, sliding down the line back onto the ledge.

As she disappeared from his sight, Decker called, "Let the peacock try! He wanted it so bad."

"Goddamn it!" Con fell to his knees looking down at her. "Get back up here *right now*! That is *an order*!" There was no response. Desperately, he added, "Lieutenant, you are disobeying a *direct order*! I'll have your stripes!"

Decker paused in her descent long enough to look back up at him. "I'm sorry, Con, but the only way you'll stop me – is to kill me," and then she vanished.

Con closed his eyes in a mixture of fury and anguish. After only a millisecond of doubt, he put his hand to the comm link in his ear and said, "Bayer, take off now. Get the wounded to *Lovelace* and...and come back for us." He then started rappelling down after Decker.

Chapter 10

Stronger Than Death

"It is love, not reason, that is stronger than death." Thomas Mann, The Magic Mountain

When Decker felt the thump of two boots next to her on the ledge, she was both surprised and surprised at her surprise. *Of course, Con would do something like this!* Without looking up from Kay's blood-stained body, she growled, "You're supposed to be on that fucking shuttle. What the hell is *wrong* with you?"

"Oh, don't you even start!" Kennedy yelled, in ascending volume, "That is *my line*, you stupid, stubborn, insubordinate—" Con didn't finish. Decker wasn't sure if it was because he couldn't bear to use the word he had in mind or if he couldn't even think of an epithet she deserved right then. When she didn't respond, he asked, "Is he still alive?"

"Barely," she admitted. "He's bleeding pretty bad from where that crab clawed him and...there's massive internal damage for sure." Decker finished wrapping him in her uniform jacket and clipped him into her rappelling harness.

She recalled the anchor from above and re-sunk it into the ground at her feet.

"Where are you taking him?"

"To the *Daleko*. The first-aid station was still intact. It's his only hope."

When she made it back down to the beach, Decker ran as fast as she'd ever run in her life but to her the journey seemed to take an eternity. She was begging, "Please, please, please hang on," as she ran, more as a litany to keep her fears at bay, than speaking to Kay himself. She jumped into the *Daleko* in a mad leap and rushed over to the first-aid station. She had bandages and the monitors on Kay and was tightening the fluid infusion wrap around his leg by the time Con joined her.

Decker took the first deep breath since Kay had gone over that ledge and wiped the tears from her face. That's when she noticed that she was covered in that pink dust. She looked up at Con and saw that he was similarly covered. "What is this shit?"

"I don't know but Pilecki said it was harmless."

"Did you see any of those crabs on the way?"

"No, but they're gonna figure out we're still here soon enough." He looked down at Kay and said softly, "I can't *believe* he survived that fall."

"When they genetically re-engineered their life-spans, they engineered in some extra-resiliency, too." She sat back on her heels and looked up at Con. "It's gotten him this far but it's not gonna get him much further. I don't know if he'll survive 'til they can get a shuttle back down here." Decker jumped up saying, "He needs to be in the Medical suite on *Lovelace*. Or something like it." She brushed some loose strands of hair away from her face, lamenting, "But how?"

"We're standing in 'how'," Kennedy said. "We've *got* to get this ship out of here."

"We tried that."

"If those goddamned crabs can do it, we can too." He rubbed a hand across his lower jaw, saying, "We just have to think. Did you try hyper-propulsion?"

"Yeah, it made it worse. The harder I fought the adhesion, the more we were sucked down." Suddenly Naiche knew the answer. She didn't know how she knew – but she did. "That's it! Down! The way out is down."

"What? That is the craziest fucking idea you've—" Con stopped cold and stared at her for a second. "You're right. That is the way out."

"Right?"

"Yeah." Shaking his head as if to clear it, Con said, "Wait a minute. How do we *know* that, Deck?"

"I don't know. I just do."

"If we're wrong, we're probably gonna die."

"I know." She glanced at the cockpit and then up at Kennedy. "Your call, Commander."

Con threw up his hands in disgust. "Oh, *now*, it's my call!" He looked from Kay back to her and then said, "Let's do it."

Decker sat in the pilot's seat and quickly donned the transdermal link. Con sat next to her and nodded once. She powered up and pointed the *Daleko* downward. They started descending through the viscid onyx sea.

Ten minutes later, Decker was wondering if it was too late to abort and head back up to the surface when the neural link suddenly indicated a slight change in the visibility of the surrounding fluid. "Is it getting lighter?"

"I think that's your imagination." Con leaned forward, closer to the view screen. "Wait...maybe not." The sea was growing lighter and gradually they found themselves sailing through clear liquid with the dark matter floating above then. "I'll be damned."

"I'm reading it as...substance unknown. Ship's not sure but at least we can see through it," she informed Con. Deck could now see huge pillars of stone all around them and she deftly guided the ship around them.

Con heaved a sigh of relief, asking, "What now?"

"I don't know – the voices in my head didn't give detailed instructions." Just when Decker was damning this plan for giving them such cruelly false hope, the neural link indicated that there might be daylight up ahead. "There's something...."

"What?"

"The way out I think." With an enormous leap of faith, Deck guided the ship into a tight dark cavern. After a few breathless minutes of intense concentration, the ship emerged under an immense open volcano-like structure. Decker closed her eyes for a second, took a deep breath and pointed the ship upward. When the *Daleko* broke above the liquid, they could see a beach area crowded with ships.

Con pointed to the black and orange stingray, saying, "Well, now we know where they're stashing the ships."

Naiche kept the *Daleko* pointed upward until they broke free of Scorpii-d's atmosphere and then plotted a rendezvous course for *Lovelace*. After hailing them and getting an okay for docking, she said, "Inform Medical to be standing by. We're bringing in a veterinary emergency. Level-one-A."

Doctor Uddin came on the line. "Lieutenant, you'll have to settle for me. Doctor Clemente's in surgery with Captain Bastié."

"Not a concern, Doctor. Please ensure all of Kay's banked tissues and serum are ready and waiting; he's in urgent need of surgery. Decker out."

They sat in relieved silence for a moment – only then did Kennedy broach the subject that hung heavy between them. "You disobeyed a direct order, Deck. That's gonna have to go in my report."

"I know." She bit her lip for a second, then looked at him. "Are you gonna have them take my stripes?"

Con rolled his eyes. "No, don't be ridiculous." He rested a hand on the cockpit control panel. With his eyes trained on his hand, rather than on her, he added, "But there will *have* to be repercussions."

"Like what?"

"I don't know." Con faced Naiche, explaining, "I'm gonna leave that up to Lindstrom. We both know I'm too close to this."

Decker leaned back in her seat and looked at the dog in the first-aid station. "As long as Kay survives, I don't care what happens to me."

Kennedy shook his head, eventually stating, "Yeah, that's the thing – I do."

***

Decker was glad to see that the decon chamber easily removed the pink dust from all of them. A delay caused by insufficient decon would have been disastrous for Kay. She gently caressed the dog laying on the transfer gurney, while saying to Con, "You know, Ricci is probably waiting outside the airlock for us."

"There's no probably about it."

"I was thinking – the fairest thing for everybody, especially Lindstrom, is to wait until *after* he's rendered his decision on my infraction to inform the captain about...about what I did."

Con seemed relieved at the idea. "You're right, that is the best thing to do."

When she saw her father a minute later, Decker immediately recognized the tightness in his jaw and the light in his eyes. He was an all-too-familiar mixture of extreme gladness and barely controlled anger. After Medical personnel had whisked Kay away, he nodded at both of them. "Commander, Lieutenant." After taking a deep breath, he snapped, "Do you want to explain *exactly* what the hell went on down there? Why you two weren't on the goddamned shuttle?!"

She pointed at the dog disappearing down the passageway, answering, "We were rescuing Kay, Captain."

"We found ourselves in a difficult predicament, sir," Kennedy said. "We knew the shuttle had to take off immediately for the sake of the wounded and to avoid another attack. When Lieutenant Decker chose to stay behind to rescue Kayatennae, I thought it best that she didn't go it alone." He exhaled briefly before adding, "The whole story will be in my report. I'll have it for Commander Lindstrom's review within the hour."

"Good." Ricci relaxed slightly and smiled at them. "Glad to see you're both okay." He clasped Naiche briefly on the arm saying, "Go ahead. I know you want to be with him. I'll catch up with you later. I've got to de-brief Pilecki now."

Decker took off in a sprint, pausing to call back, "How's Captain Bastié?"

"I'm waiting to hear. So far all signs are extremely positive."

***

The first thing Con did was go and see Aqila who immediately threw herself into his arms and kissed him passionately. She then took him by the hands and called him an absolute fucking idiot. In other words, nothing he didn't deserve. He filled her in on the quick and dirty version of the night's events and went to complete his duties.

His estimate of having his report filed in an hour was exceedingly optimistic. He re-wrote it twice, desperately trying to present the facts dispassionately. The first version had been written in anger, with him venting his still raw frustration at Decker's willfulness. He took a shower and got a bite to eat and then tackled the report again. The second draft leaned the other way, with him excusing her actions due to extreme duress and emotional strain. With the third and final version, he finally felt he'd achieved his goal of detached accuracy. Kennedy sent the report to Lindstrom's urgent attention and went to check on Kay – and Decker.

Upon entering Med-bay, Dr. Uddin directed him to one of the Intensive Care rooms. Con found Kay stretched out on the bed, with Decker lying next to him, curled protectively around his body. One hand was resting lightly on Kayatennae, the other was clutched around her moonstone locket in an obvious gesture of self-comfort.

Con looked at the array of infusion wraps and med-patches on the dog, asking, "What's his condition?"

"Touch and go. The next six hours are make or break for him."

Seeing her exhausted, disheveled condition, Con laid his hand on hers over the locket, suggesting gently, "Why don't you clean-up and get something to eat? They'll let you know immediately—"

"No. I can't leave him. I can't take that risk." Naiche looked up at him with teary eyes, stating resolutely, "I'm not going to let him die alone."

Rather than arguing with her, Con left to get the bedside chair from the adjoining room and brought it in, pushing it into place across from the existing chair. He then climbed into the makeshift bed.

"What are you doing?"

He pulled out his hand-held, answering, "I'm not going to let you *watch him die* alone." Deck nodded at him, all the thanks or acknowledgement he needed, and he settled in for the wait. Con dozed a couple of times, waking once to hear

Decker quietly chanting a haunting song in Chiricahua. "What is that?"

"In Standish it would be called...umm, *Mountain Spirit Dance.*" After a brief pause, she explained, "It's a healing song."

The next thing he knew, Dr. Clemente was shaking him awake, whispering, "Kennedy? Nils would like to speak with you, if you're up to it."

Con sat up, asking, "What time is it?"

"It's almost 0200 hours. You've been in that chair for over four hours. That can't be good for you."

Con didn't respond to that observation, thinking of all the nights he'd literally slept on rocks. He nodded at the bed, asking, "How is Kay?"

"He's still with us and getting stronger by the hour. Not out of the woods yet but for the first time, we're cautiously optimistic." She smiled at the sleeping duo saying, "Decker fell asleep twenty minutes ago. Right after he turned that corner." She shook her head in wonderment. "It's like she knew."

"It's not *like* she knew," Kennedy objected. "She did." He got up from the chair and stretched, reminding Rita, "Remember that time Decker was in here after the Burangasisti attack? Kay finally let me take him for some food and exercise, *after* she got the antidote. He *knew*. And so does she."

"I guess you're right. They have quite a connection. Never saw anything like it." The doctor amended with a laugh, "Other than you and her, that is."

At that observation, Con's face lit with a smile but it quickly faded as he remembered what it was Lindstrom wanted him for. He squared his shoulders and went to hear what his CO had to say. He looked back at Clemente, saying, "Please let her sleep a little longer. I'll be back to make sure she gets a shower and some grub." He didn't add that he hoped he wouldn't be then escorting her to a short stint in the brig.

***

Kennedy sat in the chair across from Lindstrom, watching him shake his head over the report on his screen. "I have to admit that I am a bit surprised, and *disappointed*, at the circumstances surrounding the dog's injury and rescue, Commander. I am speaking, of course, of the exchange between you and Lieutenant Decker."

"Yes, I understand, sir, but, if you'll allow me to say so, it was a moment of extreme emotional distress. When the life of someone you...care for deeply is at stake—"

Lindstrom looked up at him. "I'm even *more* surprised to hear you making excuses."

"I don't think of it as an excuse but rather as a...an explanation."

"Pa-tay-toe – pa-tah-to, Kennedy."

"Yes, sir."

"I will give you credit for writing an extremely honest and minute accounting of the events, though a reference to the exact statute which was broken would have been a nice addition. However, considering the...." Lindstrom gave a nod of reluctant acknowledgement, "...extremely stressful and emotional nature of the occasion, and the old Corps saw of, 'no lasting harm, no lasting penalty,' I think that a simple reprimand will suffice."

"Thank you, Commander. That is most generous and understanding of you."

"Yes, it is. Probably due to my exhaustion," Lindstrom said with a smirk. He then looked at Con expectantly. "I'd really like to get a few hours' sleep before the captain inevitably starts hailing me so if you'd just say the 'magic phrase', we can both get to bed. I know it's a formality but it's a necessary one."

"The magic phrase?"

"Yes," he affirmed, with a touch of irritation. "The part where you say, 'I accept your reprimand, sir'?"

"I? I accept...?" Kennedy stammered out. "*I'm* being reprimanded?!"

"Of course! Who else would we have been discussing?"

"Lieutenant Decker. The one who disobeyed my—"

"Unlawful order."

While wondering briefly if he was actually still asleep in that chair next to Kay's med-bed, Con asked, "What?"

"Commander Kennedy, you spent six years leading a squad at the front. You cannot ask me to believe that you've forgotten UDC Directive two-point-three-point-three. 'No officer shall issue an order to abandon an injured comrade unless they have proof positive that said comrade is dead'. Now seeing as the dog was still quite alive when he was brought back aboard—"

"Directive two-point-three-point-three? Look, I love Kay but we *are* talking about a dog!"

"A dog who, as we have been reminded *ad nauseum*, carries the rank of—"

"—Corpsman, third class," Con finished, comprehension finally dawning for him.

"Yes. And as such, the Directive applies to him. Now, I will withhold my opinion on the doubtful wisdom of the UDC deciding to award that rank to dogs but – they did and here we are."

Kennedy struggled for a moment for something intelligent to say but the best he could manage was, "I accept your reprimand, sir."

"Excellent, it will be noted in your service file. Dismissed. And good night. Or more properly – good morning."

***

A chastened and disheartened Con went back to round up Decker and was surprised to find Captain Ricci sitting at Kay's bedside, reading his hand-held, but no sign of Naiche. The captain looked up and said, "It's the only way I could get her to take a break. It was either this or issue an order – which I wasn't prepared to do."

After greeting him respectfully, Kennedy sagged into the chair next to Ricci and asked, "Would that have been a lawful order to Decker?"

"I would hope I still have that much authority on this ship." Matt studied him for a second before prodding, "What's wrong, Kennedy?"

"I just received a reprimand from Commander Lindstrom for issuing an unlawful order, sir. The commander will fill you in tomorrow...or later today, I mean."

Ricci nodded and pursed his lips for a second. "Well, seeing as you're here, why don't *you* fill me in – now?"

So, Con did. From start to ugly finish. He put his head in his hands and lamented, "For the first time in my career, I'm questioning my fitness for command."

In a compassionate tone, Ricci offered, "While ignorance of the rules is no excuse, you didn't realize—"

Con looked up and faced him directly. "That's the thing, Captain. I don't think it would have made any difference." Kennedy shook his head admitting, "I think I would still have issued that order on the off-chance it would have gotten Decker on that shuttle."

"And you think that makes you unfit for command?"

"Yes. I let my personal feelings for her get in the way of *doing my job.*"

"I don't think that makes you unfit for command. I think that makes you *human.*" Kennedy turned away, in no mood for platitudes. "No, listen to me. I've never seen a better Tactical team than you and Decker. You anticipate each other's moves; you read each other's minds. The other side of that coin is – situations like this."

After a second of consideration, Con said, "I guess you're right." Shedding some of his doubt and shame, he added, "You can't have one without the other."

"I wouldn't want to." When Kennedy looked at him in confusion, Matt clarified, "Like I said, we're human, we crave connection with each other." He gave a brief laugh. "It's a goddamned shame we're so *bad* at it sometimes but that's another story." Ricci leaned forward, explaining earnestly, "I know people think I'm crazy for having my daughter under my command but we spend most of our lives on this boat. All of us. And we're inevitably going to develop strong feelings for each other, blood relationship or not. Even the UDC recognizes that. It's why they allow you and Aqila to serve together, as well as me and Naiche...and the others. Hell, it's probably where Directive two-point-two-whatever came from in the first place."

The crushing weight he'd been carrying since giving that order to Decker finally slipped from his shoulders. "Thank you, Captain."

Waving off his thanks, Ricci said, "Don't mention it. Now, unless you want to take over dog-sitting duties, I suggest you get some sleep."

"If you need me to, sir, I can—"

"No, I was joking. I'm waiting to speak with Captain Bastié, anyway." Con's immediate thought was that Bastié probably needed rest much more than a debriefing. Proving Decker wasn't the only one who could occasionally read his mind, Ricci continued, "That's on *her insistence* – not mine."

"What's so important?"

"I don't know. I'll find out as soon as Clemente gives the go-ahead."

Kennedy said, "Okay then, good-night, Captain." As he stood up, he said, "If you see Decker before I do, please tell her that she has the morning off. And that she should use it to get some sleep. If necessary, that's an order – a lawful one, I believe."

Ricci nodded and gave a tired laugh. Con went off to get some sleep himself. He had a huge apology to make later.

Chapter 11

When One Door Closes

"When one door closes another always opens, but we usually look so long, so intently and so sorrowfully upon the closed door that we do not see the one that has opened."
Johann Paul Friedrich Richter

Early the next afternoon, Decker visited Med-bay and found Kayatennae sitting up and able to drink a few sips of water. Elated at his condition, she practically danced into her CO's office.

Con looked up at her approach. "Good afternoon! I hope you actually slept all this time."

"Up until about a half an hour ago, yes I did." Naiche perched on the edge of Con's desk. "Thanks for giving me the morning off."

"I'd give you the whole day but Ricci sent a message that he'd like to locate the projector on Scorpii-d and blow it to bits before we leave orbit. He intends to put a stop to their nasty little operation if we can."

"Hey, I wouldn't miss that assignment for the world! I'll grab some food and then we can take a close look at those scans from Scientific; see what we can find." She smiled fondly at him. "I was just in Med-bay; Uddin told me you and Aqila sat with Kay for a while this morning. Thanks."

"No, thanks necessary. When he wagged his tail at us, we just about burst into applause." Kennedy's gaze shifted off to the side and then back to her. "Listen, there's something I've got to say to you about what—"

"I know all about the unlawful order thing." His eyes widened in surprise so she explained, "When I got back to Kay's room, Ricci told me the whole story."

"So, you know I owe you an apology."

"You don't owe me *anything*."

"How can you say that? I threatened your stripes! I called you a stupid, stubborn, insubordinate...um...."

"Yeah, you never finished that. Besides – which of those things haven't I been on occasion?" He started to protest so she asserted, "Look, if our positions had been reversed, I'd've probably stunned you with my pulse pistol and thrown your unconscious body on that shuttle."

Con smiled, apparently recognizing some truth to that but still shook his head at her. "You're too forgiving."

"Maybe. But only with people I...uh...really care about." She rolled her eyes, saying, "It runs in my family."

"You mean Ricci?"

"Oh, hell no. I think I get the grudge holding part from him. I was thinking of my mother."

"Why do you say that?" When she didn't answer, he prodded, "Deck?"

"Ah, it's just something I found out last year."

"Which is...what?"

Slightly uncomfortable with the subject, Decker blurted it out. "They were *together* – my mother and Ricci."

Con stared at her for a second and then ventured, "Yeah...that's how it usually works...."

She tossed her head, clarifying, "*After* she came back to the UDC." When Kennedy simply nodded, she asked, "That doesn't surprise you?"

"Should it? They were in love."

"You don't find it surprising that she went right back to Ricci, just like that." She snapped her fingers. "That she didn't have some lingering resentment of him? If only on *my* behalf?"

"Oh, *that's* what's really bothering you."

"Yeah, okay, I admit it. I mean...*I* forgave him and I know he's really sorry but...in all those years they were together, he never *once* wanted to see me or even join one of her holo-chats with me. He and *shimáá* used to discuss me, I know that for a fact, and yet he never thought to himself, 'That's my kid she's talking about – maybe I should get to know her'."

"Deck, he *did*."

"Not until it was too late for *shimáá* to know about it." Decker shook her head, exclaiming, "I can't believe that never bothered her!"

"Maybe it did." Con leaned forward, advising, "Just because they were together doesn't mean she didn't have some mixed feelings. Deck, *love* is complicated. *Humans* are complicated. She could love him, admire him, *and* be disappointed in him all at the same time. You don't know all that she was feeling."

Decker had been nodding along and after musing for a short while she said, "Maybe that's what really hurts. More proof of how little I really knew her."

"You could have grown up with her in your life every day and still not know everything about her." Con gave a brief huff of amusement, adding, "Look at my mom. Dating again after all this time!"

"Oh, I think it's nice she's not alone anymore."

"Then be just as generous to *your* mother." While she was thinking that over, he urged, "Aren't you glad *she* wasn't alone those last few years? She loved him – he must have made her happy, no matter what problems they had."

Reluctantly, she sighed, "You may have a point there." Naiche threw up her hands in defeat. "You got me, again, Kennedy."

"Yeah, sure. I always win with you," he laughed.

As Decker was getting up to leave, Kennedy asked, "When you were down in Med-bay, were the captain and Lindstrom still in with Bastié? I haven't been able to get ten minutes with *either* of them today."

"No, I didn't see them. They were both in there this morning?"

"Yeah, along with Charani. And last night Ricci told me Bastié wanted to talk to him as soon as she was cleared by Medical."

"Is something up? I heard she was doing well."

"She is. That's why I'm wondering what all the conferences are about."

"I don't know," Decker said as she headed out the door, pausing to toss back, "I guess we'll find out eventually. Be back in ten."

***

Naiche grabbed one of the ready-entrees from the twenty-four-hour café and was about to head back to Kennedy's office when she noticed Tal Jacoway eating by himself, looking somewhat forlorn. On some unknown impulse she went over and parked herself across from him. "Hi," she said. "Late lunch or very early dinner?"

"Uh, hi. Late lunch. It's been a crazy day. How's your dog doing?"

"Corpsman third class Kayatennae is doing well, thank you."

He chuckled while shaking his head at her. "We're not allowed to call your dog, 'your dog', we're not allowed to call, your father, 'your father'."

"I don't make the rules," Deck said, with a shrug.

Jacoway raised an eyebrow, objecting, "Actually, I think you do."

Rather than answering, she asked, "Why has your day been so crazy? Anything to do with Bastié?"

"Yeah, how'd you know?"

"Just a hunch. So, what's up?"

"Oh, nothing much, just our entire mission to the Expanse has been scrubbed."

"What?! What happened?"

"Captain Bastié has decreed that if she can't navigate the Expanse with the QNS, no one can. End of story."

"Does she have that authority?"

"She sure does – she's mission lead."

"Okay then, her game, her call." Tal slammed his fork down and glared at her. "I'm not trying to be a smart-ass – Bastié's been there and back. She's the *only one* who has, so she's the reigning authority. Why risk your life and Zache's if—"

"And those eighty people trapped there?" Tal crossed his arms and looked away. "Just too bad for them, huh?"

"Hey," she said gently, drawing his attention back to her. "I hate abandoning them, too. I don't know what to tell you except...making it eighty-two isn't gonna help *anything*."

"There you are!" Decker looked up to see Con striding towards the table.

"Oh, yeah. I did say ten minutes. Sorry. You coulda just sent a message."

"No, I'm on my way back from Engineering. Got an urgent report from Ramsey about two of her people being attacked...." Kennedy put a hand up, belaying Decker's obvious question. "By a film of pink dust."

"*Our* pink dust?"

Before Con could answer, Tal interjected, "You have your own pink dust?"

Deck waved a hand at him, asking Con, "Did you see it? Is it that same shit?"

"It was gone by the time I got down there but it sure sounds like the same thing."

Jumping up from the table, Decker said, "Fuck! We've had a breach in decon containment?"

"I don't know. I'm headed back to my office to check the surveillance vids."

While they were leaving together, Jacoway called, "Mind if I join you? I have nothing better to do."

Kennedy looked back, answering, "Umm, sure, why not? But whata you mean, you have nothing better to do?"

"We'll explain on the way," Decker said.

***

Charani had always known that Matt could be obstinate but she was beginning to think this was stubbornness for the sake of being stubborn. "Even though she's mission lead, this is *your ship*; I know that puts you above even the QNS mission lead. You could overrule Bastié – if you wanted to."

"The question isn't whether or not I *want* to, it's whether or not I have sufficient cause." Ricci perched on the edge of his huge wooden desk and faced her. "I have no reason to believe that Jackie doesn't know what she's talking about here."

"I didn't say she didn't know what she was talking about, I said her experience may not be universal."

"Based on what?" Ricci ran his hand through his thick brown hair. "Do you know what the definition of insanity is, Zache?"

She walked a short distance away from him before she did something monumentally stupid like curse out the captain of a command ship while on that command ship. More to herself than to him, she muttered, "I must have been crazy to think I could ever do anything significant."

He answered anyway. "What the hell does that mean? Whatever happens or doesn't happen on this mission, the QNS is still a significant achievement."

"Sure, it is. A significant achievement that will be remembered as a failure. As will its designer."

"*None* of that is true. Even if you'd never invented the QNS you'd still be remembered as—"

"As what? As a UDC instructor?"

"Even if that were true – which it isn't – would that be so horrible? To be remembered fondly by scores of cadets as a beloved instructor?"

"Says the Founder's medal winner."

Ricci seemed stunned into silence for a second, then he asked, "Where the hell did that come from? If I didn't know better, I'd think you were—"

"Jealous? Yes! I'm *human* – I'm jealous of you, Matt. Who wouldn't be? They told us we couldn't have it all, but apparently that doesn't apply to *you*. Nik and I agreed that in order to be there for our kids we'd have to take a step back from our careers. And we did. And now at age fifty-one, I'm playing catch-up with my peers while you have the gold braid, your medal, and – your daughter by your side. In other words – *everything*."

Matt stared at her for a second before challenging, "Oh, you envy me? Do you want to trade places with me?" He shook a stern finger at her. "Before you answer, you *damn well* better think about what that means."

"Like what?"

He took a deep breath, looked up at the ceiling and then back at her. "When was the first time you watched Meli sleep?"

"What the hell does that—"

"No," he snapped. "I heard you out, now you're going to hear *me out*. Answer my question – when was the first time you watched your daughter sleep?"

Switching gears reluctantly, Charani answered, "Umm, that day I had the twins, of course."

"Right. Do you want to know the first time I watched *my* daughter sleep?" With a sinking feeling, Zache just nodded, encouraging him to go on. "It was in Med-bay, here on the *Lovelace*, after we almost lost her."

Totally defeated by the heartbreaking image, she sighed, "Oh, Matt, that's—"

"That wasn't even the worst part. The worst part was knowing that if she had woken up and saw me there, how unwelcome I'd have been." He bit his lip for a second, then added, brusquely, "And whose fault was that but mine?"

From experience, she knew better than to try to refute him on that, so she just murmured, "I'm sorry."

"Hey, you want more? 'Cause I've got a million of 'em. How about my memories of scouring the casualty list every fucking night, terrified that I'd find Naiche's name there but knowing that no one was even going to notify me, let alone console me, if anything happened to her."

Crushed by the realization of the burden he'd borne alone for so long, she said, "You could've told me, Matt."

He threw up his hands and got up to pace a short distance, while ranting, "You know? If there's one phrase I could happily go the rest of my life without ever hearing again – it's, 'you should've told me, Matt'."

"I didn't say you *should've* told me, I said you *could've* told me. There was no reason for you to suffer through all of that alone. My God, I was three meters away from you while you were watching them lay to rest the love of your life, and I had *no* idea. And I still don't understand why!"

Matt stopped pacing with his back to her and stayed frozen like that for a second. Finally, he turned around, admitting, "Because it wasn't until that day that the illusion came crashing down."

"What illusion?"

"Mine. My illusion, or more rightfully my *delusion*, that everything had worked out perfectly. For everybody. Naomi had her child, Gus had his granddaughter, Naiche was a happy, thriving little girl who never gave me a second thought, and...and I had made the right decision for myself ten years earlier."

"What about after that? Why not tell me then? I could have supported you. I would have been by your side throughout those trials in the Chiricahua courts, and later when Naiche was rejecting you, when you were checking those casualty lists—"

"I know, I know," he acknowledged with a resigned shake of his head. "I should have, it's just...I kept thinking that once I'd fixed my mistake, undid some of the damage, *then* I would tell people. I didn't know it would take eighteen years."

"I'm not *people,* I'm your best friend."

"You're right." Zache watched in expectant silence as he struggled for the words to explain himself. Finally, in an uncertain tone, he ventured, "I guess...when you've kept a secret for a long time, somehow *keeping it* becomes more important than the secret itself." Ricci smiled sadly at her. "So now, I'm gonna ask you again – would you trade places with me?"

"No."

"I didn't think so."

In no mood to let him win that easily, Charani countered, "Okay, your turn – would you trade places with me?"

"If I could, I'd for damned sure do some things differently—"

"That's not what I asked."

He was silent for a minute, and then with a heavy a sigh, he admitted, "I can't...say that I would." He aimed a rueful smile her way. "I guess we both got the life we wanted, Zache – we just couldn't get it without regrets."

"Most of those regrets I can live with – but letting those eighty people down...." Zache pulled her heavy hair off her shoulders with a grimace. "That's gonna haunt me forever."

"Me, too."

"Then why—"

"Because, you've got to bring me a better plan," he declared, chopping the air with his hand. "Something more than just doing the same *goddamn* thing that we've already tried and failed at. Bring me something – *anything* that's an improvement – that has a chance in hell of working, and I'll override Bastié's decision like that." He snapped his fingers to punctuate the thought.

"We don't have the luxury of time."

"Then you and Jacoway better think fast—"

His pronouncement was interrupted by VICI. "Captain Ricci, your presence is required in Lieutenant Commander Kennedy's office. Priority Situation."

Ricci pinched the bridge of his nose, as he barked, "Acknowledged." He turned to Zache, saying, with a toss of his head, "I gotta go. There's a fifty-fifty chance that has something major to do with my offspring."

"Fifty-fifty?" Zache went to follow him out of the office, adding, "I never knew what a cock-eyed optimist you are."

***

"This is the same substance that Pilecki told us about?" Ricci asked as Kennedy instructed VICI to show the security vids from the decontamination chamber when he and Decker returned from Scorpii-d with Kayatennae.

"Yes, sir. And in just a minute, you'll see that Decker and I were also covered—"

"Where's the sound?" Lindstrom asked.

"Sound isn't really necessary for what we want to show you," answered Decker.

Looking at her with a raised eyebrow, Lindstrom ordered, "VICI, playback with sound." There was an uncomfortable silence as they all listened to the recording of the Tactical officers discussing Ricci and Lindstrom's reaction to the events surrounding Kay's rescue. Lindstrom rolled his eyes at the suggestion that they wait until he'd made his decision about the infraction before informing Ricci. He took a deep breath, then observed, "That was very considerate of you, Lieutenant Decker." He held up a hand when she started to reply. "That being said, I do not require such consideration in order to do the job I was commissioned to do."

"Of course, not, sir."

"Moreover, when referring to me and Captain Ricci, it is proper to use our full titles, *even* in our absence."

Deck's shoulders drooped even lower as she said, "Yes, sir. Sorry."

"And it seems like both you and Commander Kennedy need a refresher course in the UDC Directiv—" Lindstrom stopped mid-sentence, staring at the holo-vid in shock.

"Holy fucking shit!" Ricci exclaimed as they all watched the pink dust which had been swept into the containment area under the perforated floor, ooze up and out, and then coalesce into a ball which then seemed to pass right through the decon chamber door.

Lateef leaned closer to the projection, asking, "Did the decon room lose negative pressure at any point?"

"No," Con answered. "Deck and I checked and re-checked that. Whatever that stuff is, it just fought against the airflow and...umm...crawled up. And then out the door."

"How does dust crawl?" Ricci exclaimed. "Let alone phase right through the fucking door?"

"We don't know. But we do know that stuff isn't just dust," Kennedy explained.

"It's got to be some kind of life-form," Aqila said.

"The decon sensors didn't register it as a life-form," Decker objected.

"Well, our sensors are calibrated based on the data and experiences we've had with other life-forms to date," Lateef explained. "Naturally there are going to be gaps in our knowledge base, so our extrapolations aren't perfect. Look at how we didn't recognize the Fwalayna Noctay as an intelligent species because their technology was so foreign to us."

"So, what you're saying is that I have an alien life-form with extraordinary abilities and unknown intentions running loose on this ship! Is that right?" Ricci exclaimed. "Has anyone spotted it anywhere?"

"Yes, sir." Kennedy said. "We've had a recent report of it in Engineering and while we were waiting for you and Commander Lindstrom, it was also reported in the officer's mess hall."

"Reports of it doing what?"

"It seems to appear quickly, cover a couple of people and then almost as soon as they notice it and become agitated, it...disappears."

Lindstrom asked, "What does it do to the people it...covers to cause such agitation?"

"Being suddenly drenched in a shower of pink dust seems to be sufficient cause," Kennedy answered.

“Where is it now?” barked Ricci.

“Unknown. We’re scanning for it but nothing has come up. It’s either moving so quickly that we can’t find it or sensors need to be more finely tuned.”

Lateef said, “I’ll get a team on that immediately.”

“Good.” Ricci rubbed a hand across his forehead, demanding, “Are there any reports of adverse effects from the exposure?”

“None. Captain Bastié and Commander Pilecki had the longest exposure and they’ve been thoroughly checked by Medical.”

Aqila asked, “Have the others who’ve experienced exposure been checked?”

“Kayatennae has,” Decker said.

“Besides, him,” Lindstrom said. “Who were the others and have they been checked?”

Kennedy said, “That would be me, Decker, Ensigns Levi and Ikhram from Engineering, Chef Collins and Bale Morgan, one of the line cooks. And, as far as I know, none of them have been checked since their exposure. I know I wasn’t.”

“Me, neither,” Deck confirmed.

“Okay, that’s the first thing I want rectified,” Ricci ordered. “Lateef, explain the situation to Doctor Clemente. I want you and her to go over the patient data together. See if there’s something – *anything* that the organism left behind or any physical anomalies it might have caused that were missed. Also examine these vids and see what data you can gather from them.”

“Yes, sir.”

“Kennedy, I want to be apprised of any and all further sightings or exposures. Start a data map of the results.

“Aye, Captain.”

“Okay, move, people. I’m convening a senior staff meeting in four hours to discuss results and to find a path forward." He turned to the AI unit, saying, "VICI, ship-wide alert, DEFCON-beta."

Chapter 12

Dance 'Round in a Ring

We dance 'round in a ring and suppose
But the Secret sits in the middle and knows. – Robert Frost, The Secret Sits

The senior staff listened in concerned silence as Kennedy briefed them on the results of Tactical's data mapping of the organism. "The life-form has been spotted a total of six times since it escaped from decon. There's no apparent pattern to the appearances but its MO has been consistent: it appears suddenly, covers two individuals in the vicinity, and then almost as soon as they notice it, it vanishes."

"What happens when they notice it?" Charani asked.

"It's been pretty universal," Con answered. "They freak out. Understandably."

Ricci nodded in agreement and asked Aqila, "Lateef, what can you tell us about the life-form itself?"

"A close examination of the vids revealed that it is multicellular; it moves via auto-cilia which enables it to both crawl, and float. Spectral analysis indicates that the life-form is silica-based."

"Hmm, like the Rock People," Lindstrom observed.

"Commander," Sasaki broke in, "just as an FYI, the Carraiks hate being called that – they consider it something of a slur."

Lindstrom retorted, "I'll keep that in mind for all the *many* times I interact with them, Sasaki." He looked at Aqila, saying, "Please continue, Lateef."

"That's basically it. Without a sample to examine, I can't give any information beyond that. We've attempted to gather samples from every area where it's been – from the *Daleko* to the mess hall – but there's not even a micro-trace left behind."

Ricci turned to Clemente. "Doctor, what were you able to determine from the patient data?"

"There's nothing in any med-scan that shows an after-affect from exposure to the alien life-form. With the *Lovelace* crew I was able to compare all results pre and post exposure so I'm extremely confident in my assessment."

Captain Bastié had been cleared to attend the meeting, and Ricci asked her, "Bastié, what were your initial impressions of the life-form? What made you think you were under attack?"

"Basically, the fact that it penetrated our hull. There was no other indication that it presented a threat and the ship's sensors didn't register it as one. After a few minutes I couldn't pay it any attention anyway since I was struggling to maintain a course out of the Expanse." She turned to Pilecki. "Stey, what do you remember? You were conscious the whole time – unlike me."

"Pretty much the same as you except...knowing that it's a life-form now, I wonder...."

"What?" Lindstrom said.

Pilecki looked at Clemente and asked, "Did anyone else who had exposure report feeling...or rather sensing...." He shook his head and reluctantly finished, "...thoughts not their own?"

"Why? Did *you* experience that?" Clemente asked.

"I don't know. I'm not sure," he admitted. "I thought I was just getting weird feed-back from the QNS but now...now I'm wondering if it was something else." He looked to his left asking, "Jackie, did you feel anything like that?"

"No," Bastié answered confidently. After a second of consideration, she said, "At least, I don't think so...."

Pilecki prodded, "But you're not sure?"

"It's just that...remember how I said that the QNS was fighting me about leaving the Expanse? That was weird, right?" She shrugged her shoulders saying, "But I was already experiencing severe cognitive impairment at that point so who knows?"

Clemente said, "No one who was exposed on *Lovelace* reported *anything* like that, but then again their exposures have been limited to mere minutes. What about you two?" she asked Kennedy and Decker.

"No," Decker answered. "The only thing I was thinking that whole time was how to get Kay the Medical help he needed."

"But what about knowing how to get the ship out?" Con asked.

"Oh, yeah...."

"What's that?" Ricci asked.

"Well, we were trying to figure out how to get the *Daleko* up to *Lovelace* and Deck said the way out was down through that black glue and at first I thought that was crazy but then all of a sudden I knew she was right."

"How? How did you know?" Aqila asked.

"I don't know. I just did – and Decker said the same thing." He turned to Naiche, saying, "Remember? You even said something about 'the voices in your head'?"

"Sorta. It's all a blur to me. I hadn't felt sick panic like that since...." Naiche wasn't about to admit out loud that she hadn't felt like that since being forced to kill her fellow soldiers during the war so she just trailed off, saying, "...umm, in a long time."

Ricci briskly changed the subject, asking Con, "Anything else you remember, Kennedy?"

"No. Suddenly knowing how to get the ship out is all."

"How would the dust even know something like that?" Jacoway asked. "If it can be said to *know* anything, that is."

"It might have covered those crabs when they tried to take the ship," Pilecki said. "Maybe it picked up the information from them."

"So, what does that mean?" Ricci asked. "It can read minds?" He looked at his staff saying, "We're facing an alien threat that can survive the vacuum of space, phase through a vanadinlum hull, and *read minds!?*"

"Even if it could communicate something like that," Lindstrom said. "Why would a hostile alien help you out?"

"To get up to *Lovelace*," Decker said. "Maybe it was just using us to get up here and...."

"And do what?" Clemente asked. "It really hasn't *done* anything to us. Why are we assuming it's hostile?"

"Because it keeps attacking our people?" Ramsey said.

Sasaki asked, "Is it attacking? Or is it, like the Dardanze, merely attempting to communicate?"

Bastié brushed her short dark curls out of her eyes, quipping, "Hey, buy me a drink first before you force your way into my mind." Decker burst out into a guffaw, which she quickly choked down when Ricci turned an admonishing eye from her to Bastié. "Sorry, Matt," Jackie said, giving at least a marginal impression of repentance.

Ricci nodded in acknowledgement of her apology. He then looked at Aqila, saying, "Lateef, how's the work on the sensors coming?"

"We fed the spectral analysis into the sensor data banks. That should help us ID the life-form. The only problem is, that it can evidently phase into solid objects. If it's resting inside decks or bulkheads, then I don't know if we'd pick it up."

"What if you combined the spectral data with the technology you used to penetrate the projection on Scorpii-d?"

"We could do that, but then we wouldn't get a precise location."

"An imprecise location is better than none," Ricci said.

He adjourned the meeting and dismissed everyone but before they could leave, Charani spoke up. "Before we go, Lateef, what's the current status of the Expanse?"

"The *Lovelace* is still in striking distance of it but not for long. We're continuing to detect increasing instability of the space-time field. My team is analyzing our most recent data to try to get a more precise estimate of final reconfiguration."

Since their mission to the quantum entanglement had been scrapped, no one besides Charani showed much concern at Aqila's news. Bastié seemed more interested in Zache's interest than in Lateef's pronouncement. "Why do you ask?" she inquired of Charani.

"Knowledge is power," was Zache's oblique answer, as she stood up from the table.

Bastié gave her a thoughtful look but said nothing, changing the subject entirely as she remarked to Naiche, "Lieutenant Decker, I haven't had a chance to thank you for all you did for me. Clemente tells me you not only rendered first aid, but carted me up that canyon."

"You're very welcome, sir, but it was no hardship. Nothing we didn't do every day in S and R."

With a light laugh and a wink, Bastié said to Ricci, "Self-effacing. She certainly didn't get that from *you*."

Ricci just shook his head at Jackie but Naiche was actively biting back a reply. Suddenly, Con appeared at her elbow saying, "Come on, Deck. We still have to see about finding that projector on Scorpii-d. I'll put Werther on pink dust watch."

When they were alone, walking down the passageway to Con's office, Deck said, "You didn't have to hustle me out like that. What did you think I was going to say to Bastié?"

"I don't know – but I'm sure we can file it under *nothing good.*"

***

A few frustrating hours later, Decker and Con called it a night. They hadn't been able to identify the source of the fake projection from Scorpii-d and there had been two more appearances of the alien life-form with no progress made in containing it or gathering additional information about it. Kennedy went to spend what little was left of the evening with Aqila and Decker proceeded to Med-bay to check on Kay. After finding that the dog had been moved to a regular room and was nearly ready for discharge, she stayed with him until he fell asleep. Naiche then headed to the officer's lounge for a nightcap.

After selecting a Paloma cocktail, Deck looked for a quiet place to relax and noticed Zache sitting alone at one of the tables by the view-port. She cruised on over and said, "Mind if I join you? I can see you're working on something so feel free to say no."

Charani looked up with a fond smile. "Please do. I need an excuse to stop."

After sitting down, Decker pointed at her hand-held, asking, "What were you working on?"

"A way to navigate the Expanse."

"I thought you already did that."

"A different way."

"Will it make any difference? Didn't *Captain Bastié*," she couldn't resist adding a slight edge to the name, "decree that if she can't do it, no one can?"

Zache took note of her tone and said, "I hope you're not going to hold that remark about you not being like your father, against her. She didn't mean anything by it. That was just Jackie being Jackie."

Naiche rolled her eyes, and said, "Not only was it inappropriate in that setting – but she was wrong. That wasn't a show of modesty, just honesty. I'm really *not* self-effacing at all." With a slight snicker, she said, "You can ask Lindstrom about that. If you still have doubts, wait'll you hear me bragging about sticking that landing on the canyon peaks."

"Well, that *was* impressive."

"The funny thing is, there were probably a dozen other Micro-craft pilots who could've done the same. The peacocks might've gotten all the glory but us pigeons could come through in the clutch."

Zache took a sip of her red wine and then rested her chin in her hand, asking, "You know, I've always wondered why you didn't grab the chance to be a peacock yourself."

Waving a dismissive hand, Decker answered, "With my GPA, who was gonna give me that chance?"

"Uh, Jess Coleman for one," Charani countered. "I asked her to consider recruiting you and she later told me that you turned her down flat."

Naiche was stunned silent for a moment and then exclaimed, "That was *you*? I always thought my father arranged for that."

"Why would you think that? Because he flew with Coleman?"

"No, I didn't even know that back then. It was because he'd already told me that I didn't have to go to the front, that he could get me a 'good position' somewhere else." Feeling slightly foolish, she admitted, "That's why I turned Coleman down. I was trying to prove I didn't need his help."

Zache exhaled in obvious exasperation, and then, with a shake of her head, said, "You and Matt sure have come a long way."

"You would know," Decker laughed. "After that first time you had us over to dinner. Talk about a night in 'awkward-town'."

"Was it awkward?"

"Thanks to me it was. I kept slipping and calling my father, 'Captain Ricci', and you and Nik, 'sir'. I announced that Matteo must be a very common *N'daa* name since it was shared by your son and my father rather than making the obvious leap that Teo was named after Pop."

"None of that was so bad – or your fault. Your father had been an honorary member of our family for a long time. We all knew you'd need some time to fit in."

"I guess," Decker said. "You'd think I'd be used to being the odd one out. I've been that ever since I left Chiricahua territory at seventeen. Until, the *Lovelace*, that is."

Sitting up with a smile, Zache asked, "Is that right? You feel at home here?"

"Yeah, I do. Not just because of my father either. Most of my Corps 'family' is here. There's Con, Aqila, Bly Brodie...and in some weird way I can't even explain – Lindstrom. He's the grumpy uncle I never wanted or asked for but got any way," Decker laughed.

"I'm really glad to hear that. I was worried you were mainly serving on *Lovelace* to make your father happy."

"No, not at all. Which is a good thing," Naiche said, with a rueful chuckle, "since making my father happy, isn't really my strong suit."

Zache stared at her for a second with raised eyebrows before protesting, "That's not true at all!" She mused more quietly, "Though maybe you can't even see it." Charani leaned forward, explaining, "For the longest time, there was a...a hard edge to your father. Looking back, it developed sometime after your mother died. Ever since you became part of his life – that edge is gone."

Needing some time to take that information in, Decker simply said, "Huh. Well, how 'bout that?" After a second, she added, "Thank you for telling me that. You're right – I couldn't see it."

"Glad to help."

"I wish I could help you solve your problem with the Expanse." In a desperate attempt to come up with something, Deck proposed, "What about using the *Daleko* somehow? Two ships, instead of one?"

"What do you mean?"

"I don't know...it's just...during the war, sometimes I was able to team up with another Micro-craft pilot. It made things so much easier. You couldn't take your eyes off those damn drones for a second, but with a partner, we could spell each other for a bit."

"I see what you're saying, but that's not gonna work with navigating the Expanse. It's not the same kind of 'enemy', if you will."

Decker nodded, admitting, "Yeah, I guess with the Expanse two ships would just be twice the trouble." She threw her hands up. "That's all I got."

Zache smiled and said, "It's okay. It's not your problem to solve. You've got enough on your plate with that pink dust alien running around."

After downing the last of her drink, Naiche said, "Speaking of which, I better get to bed so I can face that trouble fresh in the morning."

Zache agreed that sleep was a good idea for both of them and they left together, walking towards their quarters in a companionable silence.

***

Matt answered the chime at his quarters' door, hoping it wasn't more trouble at this late hour. He gave Jackie Bastié a relieved smile when he found her standing on the threshold. After ushering her into the sitting room, Ricci asked, "What can I do for you?"

"Offer me a drink, for starters."

"Okay, I've got whiskey, brandy, or grappa."

"I'll take a grappa, thank you."

When he brought out the bottle and two small glasses, Bastié glanced at the label, asking, "Have you lost your taste for the good stuff?"

"No, I lost the good stuff itself and restitution hasn't yet been made." At her puzzled look, he explained, "Naiche's boyfriend polished off my *riserva* grappa and she's on the hook for replacing it." He poured a glass for both of them while musing, "I need to remind her about that."

"Matt," she laughed, "you sound just like a father!"

He settled into the chair across from her declaring, "I *am* a father."

"I guess I wasn't expecting you to embrace the role so thoroughly."

"I should – I chased it long enough."

"Is that a fact?" Jackie took a sip, observing, "How very much you were keeping from me back when we were together. No wonder we didn't make it."

"Oh, it's all my fault, huh?" Before she could respond, Matt said, "Why don't you tell me what it was you were keeping from *me*? I always sensed there was something. Was it true what people said – that you were with me because I looked so much like Tony?"

"Oh my God! Is that what people said?" Bastié appeared genuinely shocked. "It's not even true, is it? Other than both of you having dark hair and green eyes, I don't see it." She sipped her drink, muttering, "As if I'd even have wanted another Tony."

"What does that mean?"

After giving him an appraising stare, Jackie put her glass down on the coffee table and leaned forward. "What I was actually keeping from you – and almost everyone else – was that Tony and I had split up right before he died. I was getting ready to divorce him." She picked up her glass, explaining, "I found out he had been cheating on me with every starry-eyed cadet and ensign who gave him a second look. Which was a lot."

"I'm sorry. That's rough." After digesting that news for a second, Ricci asked, "But why was that such a secret?"

"You remember my mother-in-law?"

"Admiral Stewart? Of course."

"Well, then you should remember that she had a lot of clout. And she said it would be a shame to besmirch the memory of a dead war hero." Jackie shrugged. "In a way she was right. We needed all the heroes we could get that year." He nodded in acknowledgment of that undeniable fact, as she continued, "And to be *extremely* honest, being known as a war widow back then wasn't hurting my career any." She smiled at Matt sadly. "What a pair we were, the fake widow, and the clandestine widower. I'll repeat myself – no wonder we didn't make it."

Though he agreed, Ricci had had enough of the subject. "I don't think this is what you came here to discuss."

"You're right. What I came here to say is that, no matter what idea Zache cooks up, you shouldn't let her and Jacoway go into the Expanse."

"You know that's she wants to do?"

"Of course! After I confronted her directly, she told me." Bastié laughed, "Unlike us, Zache isn't one for secrets. This isn't my pride or ego talking, Matteo. The QNS is a marvel but there's no pilot alive who can stand up to the Expanse. You and Zache are the only ones aboard who know that this is my last mission before I retire, so you must know how devastated I am to have my career end on such a low note. If there was *any* way of making it into a success, I'd jump at it. But I was there, and I know – there isn't one."

"I said I'd hear her out, so I have to do that. If she and Jacoway come up with something really unique – then I may not have the right to stand in their way." Ricci leaned back and ran his hand through his hair. "The thing is, this will all be moot soon. Last word I got from Astrophysics is that, with the latest shift of the Expanse, we have less than forty-eight hours before it moves beyond our reach."

"That's that, then." Jackie drained the last of her grappa but didn't seem inclined to go.

"Would you like another?"

"No," she said with a mischievous grin that warmed her honey-brown eyes. "What I'd really like, is to spend the night." While he was wrestling with his surprise, she said, "Oh, please don't try to tell me that you and Grace Stein are exclusive, because I know for a fact, that's not true."

"No, I was just wondering if you're cleared for such...activity – medically speaking that is."

"Doctor Clemente said all light activity is okay. As long as you refrain from anything really strenuous, I'm sure I'm up to it. Clemente did say a little exercise would do me good."

"Well, then...." Matt stood up and offered her his hand. "If it's doctor's orders, who am I to argue with my CMO?"

As they headed to his bedroom, Jackie said, "I'm sorry you won't be able to pull out any of your fancy moves. Especially that one where you used to hold me up against the wall."

"So am I," Matt smoothly lied.

Chapter 13

The Names of the Dead

it is prohibited to whisper the names of the dead,
as it encourages them to linger at the doorstep,
and she has already lingered, far too long – Crisosto Apache, Death

Late the next morning, an ecstatic Decker had an even more ecstatic Kayatennae released to her from Med-bay. She listened carefully to all of his discharge instructions and then got him settled into her office. She gave him the light lunch Dr. Uddin had recommended, then headed to the mess hall to get her own lunch. When she got to the table where Con, Aqila, and Bly were eating, she sensed that something was up.

After explaining that she was late due to Kay's discharge, and receiving everyone's congratulations, Decker asked, "What's the news? What'd I miss?"

"What do you mean?" Bly asked, guilt plainly written all over her freckled face.

In a voice gilded with amusement, Con explained, "Bly had some hot gossip for us."

"Well, then give, you know I love hot gossip."

Aqila warned, "You may not like this particular bit of gossip."

"Ooh, is it about me?" Deck said. "Are people once again trying to claim that I'm sleeping with Con?"

"No," said Kennedy. "Thankfully not. It's not about *you* sleeping with anyone."

"Yes, I have been rather slacking off this mission," she admitted with a smile. "Who is it?" When no one spoke up, she looked at Brodie, prodding, "Bly? Come on."

Her fair complexion flushed as she answered, "Okay, I'll tell you, but don't get mad at me. The rumor is that Bastié spent the night in the captain's quarters."

"Well, whata you know?" Decker murmured in surprise. After a second of thought, she shrugged and dug into her stew with gusto, saying, "Good for the old man. He never gets any action on *Lovelace*."

"Oh, so you *did* know they were once a couple?" Aqila asked.

Spoon frozen on the way to her mouth, Deck exclaimed, "What?! When was this?"

Aqila tilted her head in thought, before answering, "That would have been back in '22, about a year after Commander Stewart died. When I was an ensign, there was a lot of talk about her dating again...you know, because her husband had been the hero of the Battle of Rieyuu."

"No, I didn't know *any* of that. Ricci said she was an old friend but I didn't know he meant she was an *old flame*." Decker looked at Bly, asking, "Is that why you thought I'd be mad? 'Cause I didn't know they used to date?"

"No! It's because he's...." She looked around the table, silently asking if it was okay to say it, then finishing in a whisper, "*...your dad.*"

Waving a dismissive hand at her, Naiche asserted, "That's silly. I'm glad for him. Sex is good for you. Unclogs the, uh…you know…everything."

"Sex is not just something to unclog the…you know…everything," Kennedy asserted. "It can be so much more than that."

"Oh, here we go," Bly muttered. To her visible relief Tal Jacoway walked up and asked to join them.

After being invited to sit down, Tal looked around saying, "I hope I'm not interrupting, you all looked like you were deep in discussion. Is there a break-through on the pink dust?"

"No, we were talking about the restorative powers of sex," Deck said brightly. "I'm a big fan!"

After seeming momentarily at a loss for words, Jacoway asked, "Does this have anything to do with Captains Bastié and Ricci?"

Decker laughed, "Yeah, there are no secrets on *Lovelace*, are there?" When Con, Aqila, and Bly all turned and glared at her, she amended, "Okay, no secrets since my paternity was revealed." She went back to her lunch, sighing, "Geeze, get over that…."

While Tal shook his head in bemusement, Aqila, in an obvious attempt to elevate the conversation, said to Con, "*Are* we close to a breakthrough with the pink dust? Have the modified scanners helped?"

"Not really, just knowing the general vicinity of the thing doesn't do us much good. Sorry."

"That's too bad," Jacoway said. "Even if we don't go to the Expanse, I understand we're in a holding pattern until we deal with that entity."

"Yeah, Captain Ricci is unwilling to head for home with an alien life-form of unknown intentions on board," Con explained. "We'd be put under immediate quarantine if we tried to enter Uniterraen airspace."

Tal asked, "Have there been any more sightings lately?"

"Yes, three since last night. Tactical always gets there just in time to hear what we missed. If only we could figure out what it's up to. Or if only it would hang around a little longer...."

"What about when it did stick around longer?" Tal asked. "With you and Decker and before that with Bastié and Pilecki – what was unique about those times?"

Decker said, "We're way ahead of you. Both of those were on the *Daleko*, so Con and I have hung around there, repeatedly – to no avail."

"You were also calm," Aqila said. "That might be key. The entity seems to leave as soon as the people that it's...visiting get upset or agitated."

Brodie asked, "It always covers two people, right?"

"At least two," Con explained. "Several times it's been three."

With a light laugh, Naiche said, "Kinky."

Kennedy glared at her, exasperation clear in the one syllable he uttered. "Deck."

"Sorry," she said, with true contrition. "I guess I'm just a little giddy with Kay being released today."

"Oh, that's wonderful news," said Jacoway. "I hadn't heard that. How is he?"

"He's great – already itching for a job to do." She took a bite of her cornbread and then said, "You know...maybe we have one for him. We could use Kay to try to find the dust thing in its lair, so to speak."

"You think?" Con asked.

"It's harder to fool a dog's nose than any sensor in existence. That's why S and R still uses them."

"How would we give him the scent?"

"I'd take him to every place the alien's been spotted. If he's going to be able to follow it, he'll pick it up from that."

"Say you do find the place where the alien is camped out – then what?" Aqila inquired.

Shrugging, Decker admitted, "I don't know. I guess Kay and I wait – until it comes out. It covered the two of us on the *Daleko* before Con showed up so it should be willing to do so again."

"That's the best idea I've heard yet," Kennedy said. "Bring your scanner so you can record it, if it does come out." He pointed a finger at her, saying, "And remember, if it covers you – *stay calm.*"

"I've faced down death, danger, and a wall of eprions. I think I can stand up to a little dust."

With an acknowledging smile, Con offered, "I'll give you some crew evaluations to work on so you won't be bored while you wait."

"Explain to me how doing crew evaluations fixes that boredom problem?" Decker groused, but her mind was already on her planned afternoon with Kay.

***

After taking Kay to the dozen or so places that the dust alien had been spotted, Deck gave the command, "Find it!" and watched him take off. Decker then chased the eager hound around *Lovelace* for over an hour, his behavior showing he was on the trail of the alien, finding the scent, losing it, and finding it again until he finally settled down at a particular spot in the hydroponic garden and gave his "found it" bark.

There was nothing visible to the human senses, so Naiche sat cross-legged on the floor next to Kay and settled in for a long wait. She had just completed her fifth crew evaluation when she noticed the pink dust had settled on her hand. Deck looked up to find that both she and the resting dog beside her were covered with the alien substance. She stealthily turned on the recording function of her hand scanner and sent the feed to Kennedy's office while telling herself, *Calm, calm, calm. Stay calm. You're calm as a lake at midnight....*

She waited for, exactly what she wasn't sure, but was distracted when Kay caught sight of the pink dust covering his coat. He did an obvious double take and got up to shake off. The dust floated off of him momentarily but settled right back on. He then started snorting softly and twitching his ears as if the substance tickled him. Naiche couldn't help it – she laughed out loud. Her immediate thought was, *What a surprisingly pleasant sensation.*

A chill went up her spine with the subsequent realization that she hadn't thought that at all. Someone else had thought that at her – or for her. All intention of remaining calm left her as her heart started pounding so hard that the sound seemed to reverberate in her ears, a bead of sweat trickled down her forehead, and she began to hyperventilate. Naiche jumped up, screaming, "No! Get out of my head! No!"

The alien immediately disappeared but she was still breathing unsteadily and holding her head in pain when Con burst into the room, calling out, "Deck, what happened? Are you okay?" He grabbed Naiche by both arms and turned her to face him, his dark complexion ashen with concern.

Twenty minutes later, a humiliated and humbled Decker was in Med-bay being examined by Doctor Clemente with Kennedy, Lateef, and Ricci in attendance. She was there under extreme protest having tried to assure Kennedy that a visit to Medical was absolutely unnecessary as she had, for some unknown reason, completely overreacted to the alien intrusion. Con had short-circuited the argument by issuing an order, leaving her no choice but to comply.

Clemente said, "It seems to have been nothing more than a bad migraine brought on by a severe panic attack."

"See I *told* you," Deck said to Kennedy from her perch on the exam table. "I'm fine. I just acted like a complete idiot, which as far as I know, Doctor Clemente doesn't have a cure for." To cover some of her shame she reached down and petted a worried Kay, crooning, "Gave you a scare didn't I, buddy? Poor Kay."

"*Why* did you act that way?" Ricci asked. "You'd been exposed to the alien before without incident."

"Before I wasn't aware of its...presence, I guess." Heatedly, she continued, "Look, I don't know – I just freaked out, okay? Like some green ensign in their first battle." She averted her gaze from the group in a futile attempt to hide her embarrassment.

In a calm, soothing voice, Clemente asked, "What *exactly* were you thinking? What was your immediate thought after you became aware of the alien presence in your head?"

"That I didn't like it." Decker took a shaky breath and said, "It was...familiar. I mean it reminded me of something bad."

"What?" Ricci asked.

"I don't know. It was a visceral reaction." Decker didn't want to relive the memory but forced herself to edge close to it. The panic was distant enough now that she could identify the source. "Oh...it reminded me of melding with the Pakarahova."

"Have you had any more nightmares?" Con asked.

Almost simultaneously, Ricci and Clemente exclaimed, "You have nightmares about that?"

"A few. We both did," Decker explained, gesturing between herself and Kennedy. When he glared at her, she said, "You brought it up."

The doctor swiped through Decker's patient file, stating, "You never reported anything about this. When did it start?"

In a non-committal tone, Naiche answered, "Oh, it's hard to say exactly—"

"Six months ago. After they watched the surveillance vids of the melds," Aqila volunteered.

"Thanks, honey," Con muttered.

"You watched those vids!" Ricci exclaimed. "Why?"

"We thought they might have some tactical...." Decker trailed off in the face of her father's obvious skepticism. With a resigned sigh, she admitted, "Morbid curiosity, I guess."

Clemente asked Ricci, "Since the melds, have you ever had any nightmares, intrusive thoughts, anything like that?"

"No, but then I wasn't *idiotic* enough to watch the vids," Matt snapped. Naiche couldn't meet his eye until in a much softer tone he added, "And also I never had to watch my friends and loved ones transform before my eyes, either. Let alone...."

"Help them out," Deck finished for him.

"Right."

"Well, like my CO likes to say, 'forewarned is forearmed'," Decker said. "Next time I'll try—"

"There will *not* be a next time," Kennedy pronounced. "Not for you anyway. I'll take the next one."

"I think it's ill-advised for either of you to attempt contact with this alien again," Dr. Clemente said.

Con protested, "We don't know that I'll react—"

"The good doctor is right," Ricci said. "You're both out of the running for this assignment."

"I'll do it," Aqila said. When they all looked at her in mild surprise, she said, "When you think about it, this really is much more a scientific than tactical endeavor, anyway."

The captain stated, "Sounds like a plan. Get a fresh start in the morning." He looked at Decker saying, "Everybody try to get a good night's rest."

Deck said nothing, knowing how unlikely that was.

***

Matt heard the sounds of a lone basketball player before he got into the gym. He wasn't surprised that Naiche was playing at 0100 hours but he was surprised that she hadn't tried messaging him to see if he was awake, too.

He offered a quiet, "*Ya a teh,*" to alert her to his presence.

She nodded and replied, "Hey, Cap," but kept on playing.

Standing by Kay on the sidelines, he watched her make a shot from the three-point line. "Nice."

"Thanks."

Wondering if she was still suffering the sting of embarrassment, or worried about a lecture from him, he asked, "I guess you weren't in the mood for company tonight?"

To his relief, she smiled at Ricci and said, "It's always more fun to play with you than alone."

"Then why didn't you IM me?"

"Uh, I was afraid I might be...interrupting something."

He stripped off his jacket, leaving him in UDC regulation gym pants and t-shirt, while asking, "Interrupting something? Like *what*?" When she tilted her head in a reproving gesture, the penny dropped. With resigned frustration, he sighed, "Oh. I see." As he advanced onto the court he said, "I guess everyone on board has heard about...my...."

"Over-night guest? Of course."

Matt picked up a loose ball and started bouncing it in exasperated irritation. "Can't I have one goddamned moment on this ship where I'm not under constant scrutiny?"

With mock formality, Deck responded, "I'm sorry, Captain, I didn't realize this was your first week on the job." He laughed out loud and she elaborated, "Rank has its privileges but also has its pains, I suppose."

"For sure."

"Though I guess you never get used to everyone always watching you. God knows, I'm not and it's been going on since the day I showed up at The Rock." She made another shot before saying, "I sure gave 'em something to talk about today – didn't I?"

Rather than answering immediately, Ricci said, "You wanta play H-O-R-S-E or something else?"

"Let's play 'Around the Galaxy,'" she answered, naming a game where each player tried to make successful shots from nine spots on a pre-determined circuit. Deck pointed at the ball he was holding, saying, "You go first." She moved into position to grab rebounds for him.

While dribbling in preparation for his first shot, Matt said, "Since when do you worry what people think about you?"

After he made the shot, Naiche answered, "I'm not used to being thought a coward."

"No one thinks you're a coward," he insisted, moving to the second spot. "Panic attacks were not exactly a novelty during the war."

While he made the basket she said, "I know...I had a few myself as an ensign."

"Then why is this bothering you so much?" Ricci missed his third shot and turned the ball over to her.

"I feel foolish." Naiche went to the first position, explaining while she did so, "Not only because of what I did but because of *why*."

After she made the basket, Matt grabbed the rebound. He bounced the ball to her, asking, "That is the question I'm still waiting to hear the answer to, by the way. *Why* did you and Con watch those vids? 'Morbid curiosity' might have made sense right afterwards but not that much later...." She didn't contradict him but she didn't answer either so he prompted, "What really happened?"

Decker stopped playing and faced him. Ricci watched her swallow several times and waited, nerves on edge, knowing something difficult was coming. When Naiche looked up her eyes were wet but her voice was almost distant as she said, "We were talking about it one night. Con and I. About the comrade kills, I mean. He was there for all but one of them."

Matt's heart ached for her but knew there was little he could do or say to assuage the guilt she still carried about those deaths, so he just murmured, "Okay," encouraging her to go on.

"Doctor Gallego," she said, naming her Chiricahua therapist, "has told me I'm keeping their spirits trapped here with me. That I need to find a way to release them." There was a long pause but Matt could tell Naiche wasn't done. "She helped me see that the truth is – I'm angry at them. If they had just said no to the Pakarahova bonds, then...then I wouldn't have had to...do what I did." She looked down at the court, obviously ashamed of her feelings.

He knew what it was to be angry at the dead, so Matt said, "Naiche, that's a *very* understandable reaction. It's pretty common, even with ordinary grieving." She looked up and nodded sadly, silently acknowledging his point. "Why watch the vids though?"

"I think I was trying to figure out what I would have done in their place. Somehow, I thought if I could see that moment when I almost transformed – like they did, it would help me make peace with it all."

"But it didn't help," he offered.

Naiche restarted the game, moving to the second position. "Nope," she said while she took her shot. "It made everything worse." After the ball sailed in, she gave a self-deprecating laugh. "Talk about your bone-headed moves. And I made things harder for Con, in the bargain."

"Is that why you didn't tell me about these new nightmares?"

She waited until she'd made a successful shot at the third position before answering, "That and because you worried so much about the others. You worry so much about me in general."

While she moved to the fourth spot on the circuit, Matt said, "This actually wasn't a bone-headed move on your part. You were trying to fix something and you made it worse. It happens. God know I've done that plenty. And as for worrying about you, that is *my job*. Not just as your captain but particularly as your father."

Deck missed her shot and then turned to shake her head at him. "I am thirty-years old."

"Oh, is that the limit? The next time you holo-chat with your grandmother, please tell her that she's been worrying about me for twenty-years too long – and therefore, she needs to stop."

"I'm not *that* stupid," Naiche laughed, taking her place back near the hoop to snag rebounds. "If Nonna worries about you it's just because she's your mom, whereas you worry about me so much because I'm...what was that term you used, that one time? Oh, yeah. Willfully reckless."

Matt moved back to the third spot on the circuit, saying, "Yes, you sometimes make reckless decisions but those don't define you. I never meant to imply that." He made his shot and said, "Naiche, I don't worry about you because you're a *bad* daughter – I worry about you because you're *my* daughter."

She smiled at him almost shyly, confirming, "I'm not a bad daughter?"

Answering her with a smile of his own, he assured her, "No, of course not." At the fourth position he did pause to chide, "Though I am *still* waiting for that bottle of *riserva* grappa to be replaced."

"I know – you'll get it." While watching him make the basket she whined, "Why is it so goddamned expensive?"

"Because, number one, it's older than the both of us put together and number two, humankind, in its infinite wisdom, ruined most of the land the grapes grow on so it is in *extremely* limited supply." He moved to the fifth spot, and after missing the shot, proclaimed, "Perhaps you'll remember that the next time you're tempted to offer some to your very thirsty date."

Naiche grabbed the ball and headed towards the fourth position, assuring him, "Don't worry. I will *never* make that mistake again."

"See, I knew you weren't willfully reckless by nature."

Chapter 14

To Entertain Strangers

"Do not forget to entertain strangers, for by so doing some have unwittingly entertained angels." Hebrews 13:2

The next morning Kennedy and Decker went over the plan for once again finding the dust alien with Lateef. Decker was explaining how she would get Kay on the scent. "I'll take him to where we last found it in the hydroponics garden, and say, 'Find again!'. It might take him a while to locate the alien but as soon as he does, I'll IM you and Aqila and then immediately vacate the premises." She looked at Con, asking, "Good enough?"

"Yes, as long as you remember immediately means immediate—"

"Yes, sir, I have that much command of the Standish language."

Kennedy sighed at her but refrained from comment, turning to Aqila to say, "As soon as you spot the alien presence, turn on your scanner feed. Deck and I will be here waiting for it."

"Got it."

"And have your sound recorder on, too, if it's somewhere VICI can't send us the sound."

"I will."

Decker looked at Kay. "Let's go, Kayatennae. You're taking point again."

Con and Aqila watched the duo leave and then she asked, "Is Deck okay?"

"I think she's more embarrassed than anything else. She told me Ricci talked to her last night and this morning I did, too." He frowned at his wife, commenting, "She's so hard on herself, sometimes."

"All those years of being compared to her mother, and now trying to live up to her father as well. That's gonna leave a mark."

"She keeps so much bottled up, too."

Aqila smiled at him and got on her tiptoes to kiss him on the cheek, saying, "Gee, I wonder what CO she could have learned that from?"

"What? Me?" he asked, in genuine surprise.

"Yes, *you*. We're a year married and the war stories are still dribbling out. I wouldn't have even known about the Pakarahova nightmares if you hadn't kicked me awake that night. In fact, I believe almost every horror story spilled out of you in the same way – with me awakening to one of your nightmares. I think it would help you immensely if you'd just tell me about them before they got to that stage."

Con objected, "You might be right, but that shit isn't easy to talk about...." He trailed off as Aqila waved her hand at him, in a gesture of, 'exactly'.

"Honey, a lot of the memories I have are so brutal—"

"I know, but I can take it. I can share your burden. I'm not just here for the good parts of you; I'm here for *all* of you. Besides, you and Deck make a great team but sometimes you need to consult someone outside of your little echo chamber. Someone else – *anyone else* – would have immediately vetoed that idea about watching those Pakarahova surveillance vids." She shook her head, stating, "I know I would have."

"All right, you definitely have a point there." Con kissed her again, and then said, "As we say in the UDC, 'I will do better'."

"On that high note of actually getting through to you," Aqila said, "I'll be in the lab waiting for Deck's IM."

It took Kay over an hour to find the alien's presence in a remote passageway near a view-port. After being summoned by Decker, Lateef took up position; after another thirty minutes of waiting, Kennedy and Decker were alerted that Aqila was in communication with the alien. They had a good view of her sitting on a bench, covered in the pink dust, saying, "Is there a name I could call you?" After a moment of silence, she said, "Oh. Well then, is there a name I could call your...er...people?"

Con whispered into the scanner, "Aqila, you're going to have to narrate both ends of the conversation for us."

***

Several hours later the senior staff was gathered in the war room watching the vid of Lateef in communication with the dust alien. She had paused the vid to explain that the alien race was known as the Jayhine.

"So, it's not one entity?" Ricci asked.

"No, it's hundreds of trillions of separate entities combined into a community."

"Hundreds of trillions!?" Ramsey exclaimed.

"Yes. It's hard to explain, but apparently hundreds of thousands of them combine together to form the particles we can see – which they call a Jaymaelin. Each Jaymaelin 'speaks' as one while they're together but can break down into their constituent parts at any time. That's how they can pass through, what seems to us to be, solid material. Then they recombine into an entirely new Jaymaelin, always in different combinations of individuals."

Lateef restarted the vid and she could be heard explaining the human race to the Jayhine. She narrated to her colleagues, "They wanted to know how we communicate with each other. All their communication is through direct touch and they never experienced any of us touching each other. That's why they kept visiting multiple people at the same time. They were trying to catch us in the act of communicating. And as far as they could see, we didn't." She turned to Decker, saying, "Except for you and Kay. They liked you two a lot and were surprised and, I think, a little disappointed to hear that Kayatennae was a completely different species rather than some alternate form of human. They were also very sorry to have upset you."

"Yeah, no big deal," Deck said, with a wave of her hand. "That was all my fault. If you talk to them again, tell 'em I said so."

The group listened to the vid for another ten minutes where Aqila went back and forth with the Jayhine about human communication and interaction.

"How long does this go on?" Lindstrom asked.

"Quite a while, I'm afraid," Lateef responded. "They kept trying to understand why we don't communicate with other humans the way they communicate amongst themselves – and with us. They were alerted to the presence of what they thought was a species like theirs when they bumped into the *Daleko* – they sensed the neural connection between you and the ship," Lateef said to Bastié. "They were astonished to find that's not our usual form of communication."

After listening for a more few minutes, Decker said, "Damn, they really hate the concept of personal space – don't they?"

"It seems barbaric to them. They wondered how we possibly live with so much distance between us. How we ever achieve 'perfect understanding' with something as crude as verbal communication."

Sasaki broke in saying, "They've got a point, there."

After nodding in agreement, Lateef continued, "I had to explain that we don't achieve 'perfect understanding' but we do the best we can with what we've got."

"Hey, no fair," Kennedy objected. "We can achieve pretty darn close to that. We touch each other. If they could have seen us last—" When Aqila turned a death glare on her husband, he immediately stopped and amended, "Well, you know, humans *do* touch each other. Did you explain that?"

"Yes, that part is coming up. I even went into sex and reproduction."

"What did they think?" Clemente asked.

"They thought it sounded like a sad substitute for what they have together."

Laughing, Bastié said, "Not if you're doing it right, it's not."

"All right," Ricci broke in, brusquely. "Can we cut to the chase? What do they want, Lateef? Where do they come from? And how do we get them back there?"

"They come from the Expanse and all they really want, now, is to go back home."

"Too bad we can't oblige," Charani said, with a glance at Jacoway. "The Expanse is out of our reach now."

Ramsey confirmed, "Even at top speed, the best we could do is get there just in time to watch it disappear forever."

"I'm aware of that," Lateef said. "However, the Jayhine claim they know of a short-cut that would enable the

Lovelace to get to the exterior margin of the Expanse much faster. If they're right, we'll arrive well before the final collapse."

"A short-cut!" Jacoway exclaimed. "That sounds...improbable."

"I know," Lateef agreed. "I'm having those data analyzed right now to see if there's any validity to their claim."

"Wait," Kennedy said, pointing at the vid. "Are you discussing the crab-aliens with them here?"

"Yes, the Jayhine were pretty horrified by the Goaps – that's the name of the crab-aliens. They don't consider any non-Goap race as...well, worth a second thought. They look upon all foreign races as unenlightened and unfit for consideration, if you will. That's how they can kill so easily."

"Did they pick up any more information about them?" Ricci asked.

"They said that one of the Goaps who boarded the *Daleko* was worried about being sent back to the remote place if they couldn't take possession of the ship."

"The *remote place*," Kennedy and Decker repeated in unison, looking at each other with a knowing smile.

"Aqila," Con said, "can they tell us where this remote place is?"

"I'll ask."

***

Hours later, Kennedy and Decker were presenting Tactical's findings about the so-called remote place on Scorpii-d to Ricci and Lindstrom, in Ricci's office. Also present were Ramsey, Bastié, Charani, Pilecki, and Jacoway. "Once we knew a general location," Kennedy said, "it wasn't that hard to zero in on the projectors we've been looking for."

"How many projectors are we talking about?" Ricci asked.

Con pointed to the holo-schematic, explaining, "A dozen. All located in this heavily guarded mountain range."

Lindstrom asked, "Heavily guarded by what?"

"Pretty large guns, we think – located in these turrets," Decker said. "There are four to five turrets per projector."

Kennedy said, "Even with all three shuttles in action, I'm a little worried about them being agile enough to navigate the terrain, take out the projectors, *and* avoid the weapon-fire from the guns. There's a lot of tight turns in those mountains. But...we will try. The three pilots will be Jacoway, Decker, and Bayer. Deck and Jacoway will go in, guns blazing. Those two have the most battle experience and give us our best chance for success. Bayer will hang back, come in as needed, and be available to rescue anyone who takes fire and is in danger of going down."

"We can just blast the whole site from here, can't we?" Ramsey asked. "If our weapons don't have the reach, torpedoes will take 'em out for sure."

"We could," Kennedy responded. "But there are thousands of life-signs in those mountains."

"Life-signs of a race who consider all other races to be merely target practice – at best," Lindstrom reminded them.

"That's true – if we knew for a fact that they're all Goaps, which we don't," Ricci said. "Even so, I don't think that gives us license for wholesale slaughter. This isn't an 'us or them' proposition. It is more a 'preventative action'. Even if we take out the projectors, there's a good chance they'll just rebuild them."

"But that should take a good long while to accomplish, if they even accomplish it," Pilecki objected. "If we take them out now, perhaps the UDC can come back and do something permanent about this place before the Goaps can get them rebuilt. On the other hand, if we leave those projectors up, other innocent people will suffer the same fate that Jackie and I did – today, tomorrow, and maybe forever."

"Too bad the *Daleko* and *Cerxai* aren't equipped with better guns," Decker said. "This mission would be a piece of cake with those ships in action." She looked at Con, complaining, "The UDC has *got* to start biting the bullet and equipping everything that flies with battle-grade weapons. I know we like to *say* we're explorers now but—"

Bastié interjected, "Even if they had those kinds of guns, as flattered as I am by your confidence in me, Lieutenant, my neural port's been removed. And I'm pretty sure I'm not even medically cleared to fly with the transdermal link."

"Uh, sir, with all due respect, I was actually thinking of *me* as the other pilot."

Jackie glanced at Matt and said, "I stand corrected – she *does* take after you." Bastié turned back to Decker, stating, "Flying the *Daleko* up here from the surface is one thing but—"

"But Decker's actually a phenomenal QNS pilot, Captain Bastié," Jacoway volunteered.

Most of the heads in the room swiveled towards Jacoway in surprise. "And you know this, how?" Bastié demanded.

"I've seen her data on the simulator."

Zache closed her mouth, which had been agape in surprise, to ask Deck, "When were you on the QNS simulator?"

"Um, back at The Rock, in the Astronautics Lab. Captain Ricci can fill you in later. I self-reported that infraction." While Zache was giving Matt a quizzical look, Decker continued, "This is all hypothetical anyway since the ships *aren't* equipped with the right kind of guns."

"Can we wait a half a day?" Ramsey asked. "If so, Engineering can fix that."

"Can you do it that quickly?" Ricci asked.

Carla said, "Yeah, we've gotten pretty good at weapons modifications, after doing all three shuttles. I can put every free engineer to work and get it done that fast."

"Okay, then, you've got the time. Make it happen."

***

The flight teams assembled in the shuttle bay, with Ricci, Lindstrom, and Bastié there to see them off. It had been decided that Jacoway and Charani would be in the *Cerxai*, Decker and Kennedy in the *Daleko*, Bayer, Werther, and Kapoor in the L2 shuttle. Decker explained to a crestfallen Kayatennae, that he was staying behind this time. "The captain needs you to stay with him, Kay. It's a very important job." She looked at Ricci. "Isn't that right, Captain?"

"Yes, of course. Kayatennae, you're with me." The dog obeyed but was clearly not happy about it. Turning to the business at hand, Ricci said to Tal, "You're mission lead, Jacoway. You take point."

"Aye, sir." With a friendly smile and a wink, he looked at Decker, asking, "Can you deal with that, Lieutenant?"

"Of course, sir." She answered his smile, saying, "I've taken orders from *much worse* than you."

While Tal laughed, Lindstrom stepped forward and asked, "And to whom would you be referring there, Lieutenant Decker?"

Without missing a beat, Decker answered, "Lindy MacLaine, sir. My CO when I was in S and R. She was a real hard-ass."

"Is that a fact?" Lindstrom drawled. "When we get back to Uniterrae, I may have to check in with MacLaine about that."

"Do you ski, sir?"

Obviously puzzled at the non-sequitur, Lindstrom answered, "No."

"Oh, well, MacLaine took her twenty-year pension after the war ended and mustered out. She's head of Safety and Rescue at a ski resort on Centauria."

"I see." Lindstrom gave Decker a long-appraising stare and then said to Ricci, "I'll be on the bridge, Captain."

"We'll be right after you, Lindstrom." The captain announced to the assembled pilots and crews, "Okay teams, you know the mission. Take out all of those projectors but make it fast and clean. We have the element of surprise – use it. Get out of there before they can bring out their flyers. And there are to be no extreme heroics. I want everyone back in one piece. That's an order."

There was a chorus of "Yes, sir," and "Aye, Captain," and then the teams headed for their ships.

While they were settling in on the *Daleko*, Con asked Naiche, "Was that true? What you told Lindstrom about MacLaine?"

"She did retire."

"Was that who you were referring to, though?"

"As far as he knows, yeah." While Con was chortling, Deck clarified, "For the record, I was mainly thinking of Harmon." Lieutenant Commander Dwight Harmon had been Con's CO in Tactical-Front.

"Yeah, what an asshole he was."

"Still is, from what I hear. That's why I don't need Loose-Lips Lindstrom looking him up and running his mouth about what I said." Just then they got the go-ahead order from Jacoway over the comm. She smiled at Con. "Ready, Commander?'

"Let's do this, Lieutenant." She hit the power drive and they were off.

Chapter 15

Expect a Masterpiece

"When love and skill work together, expect a masterpiece."
John Ruskin

Ricci took the command chair on the bridge. He ordered Lieutenant Evans to open a channel to the comm network of the mission teams. With Lateef's modified scanners, they had a crude visual of the action on the main holo-screen. The first thing they heard was Jacoway saying, "*Cerxai* entering Scorpii-d atmosphere. Teams, check in."

"*Daleko* entering Scorpii-d atmosphere."

"L2 entering Scorpii-d atmosphere."

Jacoway ordered, "L2 take up hold position. *Daleko,* you're with me."

Decker said, "*Daleko* on your three, Commander."

"I have eyes on target number one," Tal reported. "*Daleko*?"

"Same."

"Okay, we're going in hot but steady. Decker, take out the north and east turrets first, I'm going for the projector."

"Copy that, Commander."

The only sounds they heard for a while were explosions until Jacoway shouted, "We've been spotted. I'm under fire. Get those other turrets—" More explosions, then Tal's jocular voice yelled, "Nice shooting. For a pigeon, that is." A minute later, Jacoway announced, "Projector one, target sterilized. Heading for two."

Bastié was standing next to Ricci and said, "One down, eleven to go."

"Okay, visual shows they're ready and waiting for us here," Tal shouted, his tension crackling over the comms. "Decker, we're going in Rabbit-Run," he said, describing a maneuver of random, rapid changes in a ship's flight path, presenting a notoriously hard-to-hit target. It was considered a dangerous maneuver since an unskilled pilot could easily lose control of their ship.

"Got it," Decker said, tersely.

Jacoway ordered, "Bayer, hang back and target south turret," and the shuttle took up the appropriate position.

With just a little more trouble than they had with the first projector, ten minutes later, Tal crowed, "Projector two, target sterilized. Heading for three." The bridge could hear the firing of the guns from that target before Tal even announced that they'd arrived at three. "They've spotted us. Looks like they're going for suppressive fire. Decker, I wanta go in on a Pass Weave. You know it?"

"Oh yes," she sang out. "I remember that one."

With a Pass-Weave, one of the pilots flew around the other in an intersecting and over-lapping flight path. When successful, this maneuver forced the enemy to focus on the main plane, leaving the wingman free to attack.

"Okay, good," Tal responded, "you're on wing, take out those forward-facing guns and then I'll go for the projector. Bayer, provide outside cover for both of us."

Matt leaned forward and tried to make sense of the visual. It was impossible to tell the *Cerxai* from the *Daleko* but after a few minutes it was clear that one of the ships was taking heavy fire.

In the next moment, Jacoway could be heard shouting, "That ack-ack is raining hellfire on me! Bayer, where the fuck are—"

Decker broke in saying, "Relax, Commander, I'm on your six. I got it." Seconds later the guns that had been firing at *Cerxai* were vaporized.

"Thanks, Decker, good coverage!"

Things stayed pretty much in that vein until projector seven, which proved more problematic since it was protected by an extra turret. "Bayer, I need you in tight here," Jacoway ordered. "Take out that northeast turret." The L2 shuttle managed to take out the extra target but also took a direct hit in the process.

"Bayer, what's your sitch?" Jacoway shouted.

"I'm still flying but lost my starboard stabilizer."

"Can you hang back and enact a repair?"

"Negative, the whole mount is gone."

"Okay, you're outta here. Get back to *Lovelace.*"

"I think I can keep flying if I—"

"No, you heard the captain, no heroics. Without that stabilizer, you're too vulnerable – L2 is done for the day. If we need help, we'll call for the L3. Get back to *Lovelace* and get it ready."

"Aye, sir. Good luck!"

With over half the projectors out, the projection from the planet was much weaker and they now had a clear view of the battle on Scorpii-d. However, the two QNS ships, with no back-up from the shuttle, were finding projector eight a tougher slog. One of the turrets was in an almost inaccessible location between two close mountain peaks. Decker shouted, "That west turret is a bitch. Gonna have to take out the south turret before I can even—" The west turret suddenly disappeared when the *Cerxai* came swooping in. "Damn!" Decker said. "Beautiful Bastié Roll. I'm seriously impressed, Jacoway."

Matt smiled at Jackie, saying, "Your claim to immortality," while Decker took out the south turret and Jacoway got the projector.

"Okay," Jacoway said, with palpable relief, "projector eight, target *finally* sterilized. On to nine." As the two ships headed for the next target, Jacoway teased, "Hey, Pigeon, how'd you even recognize a Bastié Roll?" The eponymous maneuver was an extra tight and demanding controlled-roll attack that only powerful fighter planes were capable of executing – and then only by the most highly skilled pilots. "I know you couldn't pull that off in a Micro-craft."

"Kelekolio didn't just teach me—"

For the first time, Kennedy's voice was heard, breaking in with the admonishment, "Deck!"

"The *basics*," Decker finished with exasperation. "Kelekolio didn't just teach me the basics."

On the bridge, Bastié asked, "What was that about?"

"Don't know," Ricci admitted.

"Kelekolio and Decker were an item at one time," Lindstrom explained, swiveling his chair to face them. When Ricci and Bastié stared at him in surprise, he said, with a touch of defensiveness, "*Everybody* knows that."

"*We* didn't," Bastié responded. Lindstrom just shrugged and turned back to the screen. Jackie leaned down and whispered to Matt, "I think I know how the entire ship found out about us."

Ricci didn't have time or interest in answering as the ships had reached the ninth projector. At this point, Jacoway and Decker were acting like pilots who'd flown together for years. There was less chatter because they were just anticipating each other's moves and while weaving in and out of the mountain range at blistering speeds, they were communicating with only partial sentences. *Lovelace* heard: "You on that—" "Already got it." and "That gun is—" "No longer a problem."

The twelfth and final projector was also covered by an extra turret but this time they didn't have the L2 to help out, and it quickly became clear that the two pilots were having trouble. Finally, they had all but two of the turrets down when Decker yelled, "We got Goap flyers coming in! Eleven o'clock!"

"I see 'em!" Jacoway answered. "That white one has the most firepower, I'll take it. Decker, get the black-and-orange one."

"I'm on it!" she said, adding in an undertone, "Hello, old friend."

"Shit! Two more coming in! Four o'clock, look like flying fortresses! I'll take 'em but you're gonna have to cover white, too, Decker!"

"Copy that!"

"Bayer, what's your status? Can you get the L3 down here?"

"On my way, Commander!"

In the meantime, Decker had gunned the *Daleko* into a half-loop followed by a half-roll, putting her well above the black-and-orange ship and head-on towards the white ship at full speed. Her first barrage clipped the forward stabilizers and put the white ship at a major disadvantage. Jacoway meanwhile, had executed another Bastié Roll, taking out one of the two massive ships.

In her next pass, Decker scored a direct hit on the white ship, taking it down, but had the black-and-orange ship on her tail; it was proving more agile and powerful than anticipated. The *Daleko* was dodging major fire and having difficulty returning it. Jacoway, who had the advantage on the other large ship, took the time to yell, as he was blasting it, "Decker! Hang on. I can help you in a minute! And the L3 is on its way."

"You worry about that other behemoth! I've still gotta a few tricks up my sleeve."

Matt could barely watch as every blast from the black-and-orange ship seemed close to taking Decker out. At one point, Jacoway ordered, "Retreat, Decker, no heroics, remember!" On *Lovelace* they could see the *Daleko* was still taking fire. "Decker, I said—"

"I heard you! I can't shake it. I'm gonna flank-flip to either get it or get the hell outta here." A flank-flip involved a dizzyingly fast, extremely tight loop combined with a steep ascent.

On-screen they could see Decker executing the maneuver just as a photon charge from the other ship exploded in her vicinity. Whether it had scored a direct hit on the *Daleko* or not, Ricci couldn't immediately tell.

The L3 shuttle swooped into the battle, heading for the black-and-orange ship but would need time to get into position.

"Decker!" Tal screamed, as he blasted away at his own target, which was in retreat. "Are you hit? Decker!"

She came over the comms, saying, "Yeah, I got a little crispy there, fire-suppression under way, but the ship thinks I got outta the way in time." The slower shuttle still wasn't in firing range.

Ricci gripped the armrests, watching the *Daleko* trailing a plume of smoke in its wake. It was now situated behind the black-and-orange ship, which was adroitly swiveling to face Decker head-on and re-gain the advantage. Before it finished its turn, the *Daleko* blasted it out of the sky.

Decker exulted, "Ship was right! We're fine. More than fine – now."

"Good," Jacoway answered, as he finished off the last Goap ship. "Flyers are all dust. We gotta get those last two turrets quick and get outta here before more show up. Suggestions?"

"We take opposite sides. You go high; I go low. Give it everything you've got. Just don't hit me. In the meantime, Bayer can get the projector."

"You trust me that much, Decker?"

"Hey, you're the one who's gonna have the transdermal link pilot firing right at you."

"Let's do it."

On screen it did look like the *Cerxai* and *Daleko* were firing directly at each other; Ricci had to force himself not to close his eyes during the maneuver. But he was pretty sure he didn't breathe until Jacoway announced, "Twelfth and final target sterilized. We're on our way home, *Lovelace*."

Ricci hit the comm button on his chair. "*Lovelace* here. Good job, mission team. We'll see you soon."

Comms were still open and Tal could be heard asking, "What was that move you made there with that black-and-orange ship, Decker? A flank-flip?"

"Yeah, something us pigeons cooked up at the front. You combine a tight loop with an immediate vertical climb. Not only makes you almost impossible to hit, but if you time it right, it puts you right behind your attacker."

"Neat trick! Why don't we teach that in Astro?"

"*Because*," Charani answered, coming over the comms for the first time, "if you don't execute it perfectly, you can choke off your power rod and stall out."

"You never stalled out, Decker?"

Her first answer was a chuckle, then she said, "Wanta field that one, Con?"

"Let's just say, she didn't figure out how to restart a Micro-craft in free fall to win bar bets."

"Winning bar bets was just a bonus!" Decker announced.

On the bridge, Jackie looked at Matt and asked, "Did your mother ever lay a curse on you about having a child just like you?"

"Of course."

"Tell her it worked."

***

Late that afternoon Ricci had the *Lovelace* at DEFCON-gamma just in case the Goaps decided to launch an attack in retaliation for the destruction of their projectors. However, no overt response was forthcoming before they left the vicinity of Scorpii-d, and the crew turned their attention to the next order of business – getting the Jayhine home. Lateef and Ramsey had consulted together over the reputed short-cut that the Jayhine knew to the Expanse. The senior staff gathered in the war room to discuss their findings.

Commander Ramsey was saying, "The idea of dropping out of L-speed mid-cosmic string, while *theoretically* possible is certainly...unconventional."

"What risks are we taking if we do so?" Ricci asked.

"Burning a lot more fuel than I'd anticipated and straining the engines a bit, is about all."

"Could you quantify 'a bit' for me?"

"Nothing they can't take."

While nodding thoughtfully, Matt asked, "What would our fuel situation be like afterwards?"

"We'd still have plenty to get home – though we'd be going a tiny bit slower than I'd planned."

Ricci leaned back in his chair and ran a hand across his chin. "How did they say they know about this short-cut, Lateef?"

"It wasn't really clear. My general understanding is that they have a strong homing instinct and possess extensive understanding of the physics of the Expanse. Certainly, beyond what we know."

"Well, that's not hard – is it?" Charani said. "Knowing *anything* for sure puts them a leg up on us. At this point, anyway."

"Say this short-cut works – that just gets them *to* the Expanse," Bastié said. "Since we're unable to enter it again, how do you propose getting them *in*?"

"I've discussed with Tactical the possibility of removing the photonic payload from one of the torpedoes," Aqila explained with, a glance at Kennedy. "We would then use that as a...delivery mechanism."

Lindstrom asked, "So you're proposing shooting them in?"

"Basically, yes."

"This plan gets more *interesting,"* he responded, raising an eyebrow, "with every added nuance."

"We do have an obligation of sorts to get them home," Clemente said. "However unknowingly, a UDC vessel did carry them out."

Kennedy added, "And they did give us information that helped us take out those projectors."

"All good points," Ricci said. "I'm strongly inclined to greenlight this. What say you, Commander?" he asked Lindstrom.

"Oh, why not?" Nils answered, throwing up a hand. "Taking directions from sentient pixie dust was the last open square on my bingo card of space weirdness."

Matt snorted with amusement. "With that ringing endorsement, okay, we're in. Let's put this plan in motion." He said to Ramsey, "Commander, plot a course for the Expanse with Petrović. How long do you think it will take to prepare?"

"I want to calculate all the parameters with extreme care, and then we'll have to recalibrate the Nav-sat...give the engines a thorough once over...so give us...twenty-four hours?"

"Consider it done." He turned to Lateef and Kennedy, ordering, "In the meantime, Tactical can assist Scientific in getting the Jayhine's ride ready."

While the staff was leaving the room, Charani asked Bastié, "Are you free for dinner?"

When Jackie smiled coyly at Ricci before answering, "Actually, no, I already have plans for tonight. Why?" Matt saw the knowing looks pass between many of his crew. He assiduously ignored the undercurrent and avoided appearing too interested in what Bastié had to say.

"I just wanted a private discussion – that's all."

Ramsey volunteered, "If you want, Charani, you can use my office right now. I'll be down in Engineering for a few hours at least."

Zache thanked the chief engineer and went off with Bastié. Matt didn't give it a second thought until he saw that Pilecki and Jacoway were also heading for Ramsey's office. He couldn't help but wonder if there was a last-minute plan afoot for a QNS vessel to enter the Expanse. Ricci mentally shrugged, figuring he'd deal with it when and if it was presented to him.

Chapter 16

Times of Challenge

"The ultimate measure of a man is not where he stands in moments of comfort and convenience, but where he stands at times of challenge and controversy." Martin Luther King, Jr.

"Will you at least listen to what I have to say?" Charani pleaded.

Bastié insisted, "Zache, I'm telling you – I was *there*, and I know there's no way for a pilot—"

"What I'm talking about is using two pilots!"

"How does that help?"

"I read your mission report. You felt fatigue within the first ten to fifteen minutes. What if, at that juncture, you'd been able to take a break and let Stey take over?"

Pilecki objected, "I'm no pilot. And without a port—"

Charani turned to him, "What if you'd *had* a port? Wouldn't you been able to spell Jackie, in that case?"

"Frankly – no. After watching Jackie navigate the Expanse, I can tell you – that's not the time or place for an amateur. Even a proficient one."

"Okay, but we're not talking about you and Jackie, really. What if it were me and Tal? Is there any reason the dual pilot idea wouldn't work?"

"In the first really obvious place – you don't *have* a port and in the second, the QNS wasn't designed for two pilots," Pilecki answered. "It can't link to more than one pilot at a time."

"I can get a port inserted right here on *Lovelace*. It's a slightly demanding medical procedure but doable. I asked Clemente – she said after the experience of removing Jackie's port, she would have no problem installing one. And to your second point – it's not two pilots at the *same time*. It's two pilots taking turns – spelling each other so that the strain is shared. If the two pilots alternated back-and-forth, back-and-forth, neither should get too fatigued and over-extended. Right?"

"I don't know...I just don't know," Bastié murmured. She looked at Zache, shaking her head in bemusement. "Where did you even get such a wild idea in the first place?"

"To tell the truth, it was from Decker."

Jacoway asked in surprise, "She suggested doing this?"

"Not precisely. She suggested using two ships. She told me how occasionally Micro-craft pilots used to team up during the war and share the burden of fighting a drone – how much easier it was. Now, two ships wouldn't work in the Expanse – but two pilots in one ship, ah, there you have the same idea of sharing the burden in a way that works."

"I believe that we still have the same problem with you and Tal as with me and Jackie," Stey asserted. "Zache, I love you like a sister and you're a wizard with astronavigational theory, but – you're no pilot. You've always said so yourself."

"I can do this," she insisted. "I've got to at least try. We owe it to the eighty people there in the Expanse *to try*."

Shaking her head slowly in deep thought, Bastié mused, "Even if I sign off on this – you'll still have to get Matt's approval as well." She looked at Zache, asking, "You think Ricci's gonna let his best friend go into the Expanse on a wing and prayer?"

"*Will* I need his approval though? *You're* mission lead; your say-so should be good enough."

"Okay, he won't have to approve it but he could very well put a stop to it. If he had sufficient cause."

"He doesn't though. I admit I'm no pilot but I've done a couple dozen test runs on the simulator. That experience counts for *something*."

Jackie looked at Tal. "Your life would be on the line, too. Be honest, you've seen Zache's data – could she do this?"

"No." Jacoway looked at Charani, shaking his head. "I'm sorry, Zache, you're not the right person for this mission. But—"

Charani threw up her hands in despair. Interrupting him in frustration, she said, "Well, that's that—"

"Decker is," Tal continued.

"What?!" Bastié exclaimed. "She's not even *part* of this mission."

"Come on," Jacoway insisted. "You saw her in action over Scorpii-d. And that was with a transdermal link. She could do this with me – in fact, she's the *only one* on this ship right now who can."

Bastié said, "Now, *that* is something Matt would have to sign off on." She looked at Charani, asking, "What do you think?"

After expelling a long sigh full of uneasiness, Zache finally said, "I think he always says that on this ship he's her captain first and her father second."

"Wow, talk about putting that claim to the test," Stey said. "But we're getting ahead of ourselves – we don't even know if Decker would agree to take this on. What we're proposing is definitely dangerous and, let's face it, a little bit crazy."

"And at this point unauthorized," Bastié added.

"Great," Tal said. "Dangerous, crazy, and unauthorized – those are her three favorite things." After exchanging a smile with Zache, he said, "I'll go talk to Decker about it."

"No, let me do it," Charani insisted.

"Why?"

"*Because*, Matt is much less likely to throw me out an air-lock when he hears that I was the one to propose this." She looked at Bastié. "Don't mention anything about this to him over, uh, *dinner*."

"Don't worry, I'm not even tempted!"

***

Early the next morning Ricci was sitting in his office listening to the proposal from Decker, Charani, and Jacoway. He was trying his best to appear calm but was having trouble containing his mounting fury. "Okay, I have all the information I need to render a decision. Jacoway, Decker, you're dismissed. Charani, please stay behind."

Rising from his chair, Jacoway said, "I'm sorry, Captain, but did I miss your decision?"

"No. I said I had all the information I required. I did not say that I had made one."

"Yes, sir."

Naiche hadn't yet budged, so Matt looked at her and commanded, "What part of 'you're dismissed' didn't you understand, Lieutenant?"

Father and daughter stared at each other for a long second, then Decker said, "Yes, sir," and left with Jacoway.

When the door closed behind them, Zache said, "Now I know how all those cadets felt when I asked them to see me after class."

Rather than responding to her quip, Ricci stood up and paced the room for a few seconds, trying desperately to contain his temper. The best he could do was to snap at Zache, "How *dare* you go behind my back and ensnare my daughter in this reckless plan!"

"I didn't ensnare *your daughter* – I recruited your lieutenant." She stood up to face him directly. "My God, Matt! You make it sound liked I lured her away with some candy. She's not a child."

"She's *my* child."

"Not on this ship, she's not. At least that's what you've always claimed. Captain first, father second – remember?"

"You're gonna throw that in my face, huh? I would never do this to you! I would *never* try to recruit one of your kids—"

"Damn right you wouldn't! Because my kids aren't in the UDC. Yours is. She's a first lieutenant who took an oath to serve *as needed*, not as convenient. And you love having her serving with you – don't try to tell me you don't. Well, this situation is the reverse of that medal."

"She has risked enough, suffered enough, sacrificed *enough*!" he shouted. More calmly he added, "She's already given more than enough to the UDC for one lifetime."

Ricci thought he'd scored a point when Charani sagged back for a moment – apparently unable to refute his statement. But then she looked up at him, saying sadly, "You're right – she has. But it doesn't work that way. You of all people must know that. If there was *anyone* else—"

"Did you even think about using someone else first?"

"Yes! I thought of me," she retorted, pointing at herself. "*I* wanted to be the second pilot on this mission but Jacoway refused to consider it. He won't fly with me on this one. He says Decker isn't just the best co-pilot for this mission – she's the *only* one."

Matt walked away without responding to her directly, muttering, "This is the Tripoli mission all over again." He waited for the inevitable question but it didn't come. Of course not. At this point, everyone in the UDC knew the codename for the mission that had ended in Naomi Decker's death. It was on her memorial for God's sake.

"If I thought, even for one second, that was true—"

Spinning around to face her, he insisted, "Well, I don't think it – I know it." Ricci stared her down coldly, vowing, "But this time it's different. This time it's within my power to stop it."

"Yeah, you have the power," she admitted. Defiantly she added, "But, *Captain*, do you have the *right*?"

Stung to his core, Ricci had no immediate response. He turned to leave, saying, "If you'll excuse me, Commander," while indicating she should proceed him out.

"Where are you going?"

"If you must know, to consult with my first officer. I do have *that right* – just for your information."

To Ricci's frustration, Lindstrom had been meeting with Ramsey when he went to find him. He left word with VICI that Lindstrom should report to the captain's office as soon as he was free. Matt was trying – and failing – to focus on Ramsey's report on the Nav-sat recalibrations when Nils chimed at the door.

After Ricci explained the situation to him, Lindstrom sat silent, shaking his head in disapproval. Finally, he leaned forward to say, "That is the craziest idea we've heard yet on this mission – and that's saying something!"

"So, you think I should veto the plan?"

"Well, no, I can't go quite that far. We've definitely approved crazier ideas on *Lovelace* over the years."

Trying to keep the surprise out of his voice, Ricci confirmed, "You're recommending approval?"

"What I'm recommending, Matt, is that you ask yourself one question: 'What would I be doing if the proposed co-pilot on this mission was Petrović rather than Decker?'" Nils looked at him expectantly.

"That is the sixty-four billion unno question – isn't it?" Ricci ran his hands from his forehead into his hair while giving it some thought. As much as it pained him, he finally had to admit, "If it was Petrović, I think I'd...*reluctantly* approve it."

"You have your answer."

"God*damn* it," he cursed softly to himself.

"That being said," Lindstrom stated firmly, "while Decker's captain might have to approve this mission, there's nothing preventing *her father* from trying to talk her out of taking on what is, after all, an entirely voluntary role."

With a sad smile, Ricci asked, "And what do you think her father's chance of success is, in that regard?"

"Admittedly slim. She can be so impetuous and stubborn." Though he didn't like to hear it stated by someone else, Ricci couldn't object – not until Lindstrom mused, "If only she had inherited your looks and her mother's personality rather than the other way around...."

"Excuse me, Commander?" Matt demanded.

"Okay, granted, she probably wouldn't be anywhere near as good looking—" Nils stopped dead, finally taking note of his captain's ire. "Never mind...forget I said anything, sir."

"No, please continue, Lindstrom. I want to see if you can somehow produce a third foot to *stick in your mouth*."

Rather than responding directly, Lindstrom said, "If there's nothing else, Captain?"

Ricci dismissed him and contemplated how best to dissuade Naiche. After a minute of thought, he barked at the AI unit on his desk, "VICI, arrange for a private lunch in my quarters for two. Then inform Lieutenant Decker that she's having lunch with me – in my quarters."

***

Naiche smiled at Con when he bounded into her office at 1120 hours. Before he could say anything, she said, "I can't have lunch with you today."

"Busy getting ready for your big mission?"

"Busy having lunch in the captain's quarters."

"Whose idea was that?"

"Guess."

"Do you think this means he's denying your participation and wants to explain himself?"

"Either that, or he's trying to talk me out of this. Could go either way." She looked at the chrono. "I'll find out in ten minutes."

"I'd wish you luck, but in the case of this mission – I'm not really sure what that would mean."

"I'll come through this all right. I've been up against far worse than this."

"That's for sure." Con mustered up a brave smile, offering, "And, of course, you'll come through okay. That's an order – a lawful order."

She laughed, then bit her lip, finally venturing, "But, you know...I mean, just in case...." Decker drew a deep breath, then finished softly, "...take care of the old man for me?"

Kennedy's smiled dimmed but he promised resolutely, "I will. I promise."

Decker stood up, saying, "Okay, I better go." She looked at the dog chewing a rubberite bone on his bed in the corner. "It's not the mess hall so I see no reason why you can't come, too, Kay. Let's go."

After Ricci welcomed her into his quarters, she asked, "So...to what do I owe the honor?"

"What honor?"

Naiche watched Kay settle himself on one of the area rugs and responded, "Private lunch in the captain's quarters."

"You're not having lunch with the captain." He smiled warmly at her, explaining, "You're having lunch with your father."

"Oh, I see."

"Do you?" Ricci waved his hand at the table, where lunch was waiting. "You like trout, right? I mean, even if you're not the one who caught them."

"I love it no matter where they came from." Decker took her seat at the table where a platter containing two beautiful grilled trout was on display. "Fresh out of the tank, huh?"

"Collins chose them herself."

"Captain—"

"Not for this meal."

"Okay, then, Pop, please tell me what this is all about."

"Let's eat first, then we'll discuss the serious stuff."

Though she wanted to insist he tell her what his decision regarding the mission was, Decker realized that would be an unwise tactic with her captain – *or* her father. "Okay, since we're avoiding the obvious subject, what do you want to talk about?"

As he helped himself to the green beans, Ricci asked, "Is there any gossip floating around the ship that I don't know about? Meaning something Lindstrom hasn't heard."

She thought about it for a second. "There's the persistent rumor that Lindstrom and Clemente are finally gonna move in together – or at least share quarters on *Lovelace*."

"Never gonna happen; Rita's too smart for that."

"Bly and Bayer broke up – *again*."

"How many times is that?"

"Everyone's lost count – including Bly and Evelyn." His laughter was abruptly cut off when she added, "Oh, and people are taking bets on whether or not I'm gonna get a new mommy."

He glared at her a second before pronouncing, "Not funny."

"Hey, I'm just the reporter here." She took a gulp of ice water and asked, "Why didn't you tell me about you and Bastié before this?"

"It was all such ancient history. I rather expected it would stay there." Ricci then launched into the story of him and Jackie Bastié. Between that and Decker reporting on Kay's continued recovery, they managed to avoid the subject of the Expanse right up to when Ricci said, "Let's have our coffee in the sitting area."

Naiche sat on the edge of the loveseat awaiting his pronouncement. She was certain that this whole production had been an effort to soften the blow of his proscription against her participation. Therefore, she was startled when Ricci said, "I have come to the realization that I have no alternative but to sanction your role in the Expanse rescue effort." Before she could thank him, he put a hand up to stay her. "That's what Captain Ricci has to declare." He leaned forward. "What Matteo Ricci wants to ask, is that you not do this."

"Pop, I have to."

"No – you don't."

"I'm not sure I could live with myself if I didn't."

"Let's find out. I'm betting you live a lot longer this way."

Tilting her head in an imploring gesture, she chided, "Have a little faith in me."

"Believe me, it's not *you* I don't have faith in."

Naiche rested her hands on her knees and stared into her father's eyes. "Listen to me – I *can* do this. I *know* I can. But either way, I've got to try. Too many lives hang in the balance."

Ricci jumped up from the chair, muttering angrily, "I don't believe this...I don't fucking believe this!"

She stared at his back, stunned and bewildered. "What? What don't you believe?"

"That's almost word-for-word what she said to me." A cold shudder ran through Decker as Ricci turned and gave the unnecessary explanation, "Your mother, that's what she said to me the night before she—" He stopped abruptly, eyes closed in pain. He was finally able to force it out, finishing, "The night before she was murdered by the Eternals."

Naiche stood to face him, confirming, "You were with *shimáá* that night?" When he nodded, Decker asked, "You tried to talk her out of negotiating the truce? Why?"

"Because *I knew* – I knew it was a goddamned disaster in the making. I knew the chances of success were so fucking slim, that it was an indefensible ask on the part of the Admiralty." In a far-away voice he finished, "But ask they did, and she thought it her duty to try." Matt drew a shaky breath, pleading, "Your mother wouldn't listen to me. *Please* don't make me go oh-for-two."

Never before had Naiche felt so torn. She walked over and put both hands on her father's arm. "I'd give anything to be able to say yes to you. I really would – but I can't. But I also promise you – this isn't the same thing as the Tripoli mission. *It's not*. We're not dealing with the Eternals here. And Jacoway and I make an unbeatable team. You've *got* to believe me."

Ricci stared at her in silence for a few seconds and then nodded. "Okay, then...I believe you." He looked off to the side and then back at Naiche, saying, "I guess you've got an appointment in Med-bay right now – to get a hole drilled in your head."

He managed a small laugh when Decker said, "Yeah, considering how hard my skull is, Clemente has her work cut out for her."

Chapter 17

The Storm Terrible

"The fishermen know that the sea is dangerous and the storm terrible, but they have never found these dangers sufficient reason for remaining ashore." Vincent Van Gogh

The short-cut had proved to be everything the Jayhine had promised. The *Lovelace* dropped out of L-speed mid-string to find that the Expanse was no more than a few astronomical units away. En route, Decker had stowed her gear aboard the *Cerxai* and went to work trying out her new neural port on its simulator. The time had flown by and she was surprised when VICI informed her that they had arrived. She climbed out of the *Cerxai* to find the shuttle bay buzzing with people engaged in last-minute preparations.

Decker would have preferred a quiet send-off that would have allowed a private word with both Ricci and Kennedy, but it was not to be. The shuttle bay was filled with mission personnel, the command crew, and well-wishers from every department on the ship.

Once again, Kayatennae was disappointed to be staying behind but Naiche bent down for an earnest "conversation" with the dog, assuring him that the captain needed him and she'd be back soon. When she straightened up, she looked into her father's worried eyes and wished she wasn't the cause of it.

Con thumped her on the shoulder and said, "I know you'll make us proud." Deck smiled, and promised she'd try. Before stepping back, he offered, "*Ka dish day.*"

She returned the farewell and turned towards her father. "It's gonna be okay, Captain." Decker gestured to Jacoway. "We got this."

"I know," Ricci answered, his strong, assured tone asserting a confidence she knew in her heart wasn't quite authentic.

Tal finished receiving some last-minute instructions and reminders from Ramsey, Bastié, and Charani and walked over to where she was standing. "Ready, Lieutenant?"

"Let's do this, Commander."

They both saluted Ricci and boarded their ship. Jacoway explained how they were going to flip the QNS channel back and forth between her port and his. "Once we're in the Expanse, we'll immediately start alternating pilot control. I'll set it on auto, with twelve minutes on, twelve minutes off for each of us. If that doesn't work for one or both of us, we'll adjust as needed. Until we get into the Expanse, I'll be doing all of the piloting."

"Got it." She glanced at the pink ball of dust behind them, promising, "We're going to get you home in a few minutes, folks."

The *Cerxai* took off and Tal paused the ship at the entrance to the spatial phenomenon which to the human eye looked like just so much empty space. Decker glanced down at the data screen where she could see the clear demarcation between their present location and the Expanse. Tal pointed at the screen, saying, “Looks like there’s a cosmic pool right near our entrance point.”

Decker had been briefed on the subject of cosmic pools, which were tangled whirlpools of cosmic strings that littered the Expanse. Dr. Okeke had posited that these phenomena held the quantum entanglement together. The pilots had been strongly cautioned to avoid the pools at all costs. No one knew what happened to a ship that became ensnared in one but Deck supposed, as Con would say, it could be filed under “nothing good”.

Jacoway hailed *Lovelace* and asked to be patched through to Engineering. “*Cerxai* here. Initiating tether sequence.” After Ramsey had confirmed the return of the virtual tether, he said, “Okay, we’re going in.”

Decker could immediately tell that Tal’s entire concentration was focused on piloting the ship – she watched his hands moving rapidly over the control panel, watched the screen’s display shifting at a furious pace – indicating the second-by-second changes necessary to keep up with the fluctuating space-time field they were traveling in. She held her breath, awaiting her turn; when it came and the ship flipped over to her control, she was bombarded with in-puts from the navigation system. Her hands flew over the controls, her body tense with the effort required, the strain was just starting to get to her when control flipped back to Jacoway.

After breathing an enormous sigh of relief, Decker looked back to check on the Jayhine – they were gone. She wondered at what point they’d left the ship but didn’t have time to think about it for long since she was responsible for checking the in-put sensors for any signs of the *Burnell* or the *Meitner*. There were none. With the tether, communication with *Lovelace* was strong and steady despite the quantum entanglement so she checked in with them. Evans responded to her hails and Decker informed the bridge crew that they were in the Expanse, and thus far the dual piloting set-up was working well. Then it was her turn at the controls again.

Each time Jacoway took over for her, Decker's relief was stronger. She could now appreciate what Bastié had said about the Expanse and agreed that the strain of piloting under these circumstances was more than any one pilot could bear. She and Tal had no real opportunity to discuss anything but they were still working together in perfect harmony.

They'd been in the Expanse for three hours when, during one of her "breaks", Decker's heart raced upon detecting a distress signal. After a short scan of the sensor logs, she was able to confirm – it was from the *Meitner*! She announced the discovery aloud even though she knew it unlikely that Tal could respond, then she hailed the other ship. "*Meitner*, come in. This is the UDC rescue vessel, *Cerxai. Meitner,* do you copy?"

A joyful voice responded, "*Cerxai*, we copy! Holy hell, who are you and how did you find us?"

"The how is a long, long story. I'm Lieutenant Naiche Decker and I'm going be sending over a set of directions for getting you out of here. Who am I speaking with?"

"Lieutenant Tyra Greenleigh."

"Okay, Greenleigh, get these to your Engineering head ASAP." Decker hit send on the packet containing instructions for the *Meitner* to follow the tether to the *Lovelace*. Then it was her turn to take control.

When her break came, Tal took a moment to inform her that the captain of the *Meitner* wanted to talk to her. Decker smiled broadly since she had an idea what he wanted. "Decker, here. Am I speaking to Captain Gabe Mazuzan?"

"Damn right you are, Decker. How's that dog you stole from me?"

Naiche laughed since Mazuzan had been the head of the UDC Search and Rescue Corps when she left and was the one who had signed off on her taking Kayatennae with her. "Sir, I do believe that was your signature on his release orders."

"Ah, like I had a choice – that fool dog was besotted with you and wasn't gonna work with anyone else." He dropped the mock gruffness and asked, "Is he there on the ship with you now?"

"No, sir, he's on the *Lovelace* and he's doing great. In a way he's partially responsible for this rescue, so he's still doing the job he was trained for."

"What'd he do? Sniff out the Expanse?"

"In a way. I'll fill you in later. Is the *Meitner* on the tether yet?"

"We're just about ready. Another five minutes engineering tells me."

"Great. Have you had any contact with the *Burnell*?"

"As I was telling Jacoway, only once since we've been here, and that was early on. The instability of the space-time field is getting much worse."

"We know; the whole field is collapsing. We're facing a major shift soon. How's your crew holding up?"

"I don't mind telling you – we've spent the last few weeks on the knife edge of despair. We've been conserving fuel and rationing food for the past month. I was beginning to think we were just gonna have to find a nice planet and start the human race all over again."

"Glad it didn't come to that. I've got to go now, sir, we'll catch up later." Naiche took her turn at the controls, buoyed by their success with the *Meitner*. She and Jacoway both needed the lift since they were feeling the strain of piloting the *Cerxai*. Deck was trying to ignore the tightness in her shoulders and the pounding in her head and even though she knew it was exactly twelve minutes each time, her shifts seemed to be getting longer and longer.

Ninety minutes later they'd gotten the word that *Meitner* had made it out of the Expanse. Despite his obvious stress and weariness, Tal glanced at her, his eyes shining with delight. "That's twenty-five lives we just saved."

She smiled back. "Twenty-five down, fifty-five to go."

***

Ricci watched the *Meitner* emerge from the Expanse with a jumbled mix of emotions: pride at the accomplishment Decker was a major part of, relief that not only was one of the ships safe but that they'd proved the concept worked, and lastly: a selfish wish that Naiche's part in all of this was already over. He checked in with Mazuzan to ask about the state of his crew and ship. It sounded like they'd had a grim few weeks but were in good enough shape to make the journey to Uniterrae.

"You sure you don't require anything from us right now? Some food? Or Medical supplies?" Ricci asked, knowing that the *Meitner* was a small vessel and not as well equipped as *Lovelace* or even the *Burnell.*

"Not right now," Gabe replied. "We're gonna catch our breath, maybe do a little bit of celebrating, and wait for the *Burnell* to get the hell out of there. We gonna caravan it back home?"

Ricci confirmed, "That's probably the wisest measure. We'll be in touch, Captain. *Lovelace* out." He looked at the chrono; Decker and Jacoway had been in the Expanse for going on five hours. "Evans, hail *Cerxai* and ask for a status update."

Evans complied and a few minutes later Jacoway's voice came over the comm. "Jacoway here. We're hanging in there, *Lovelace.* Still looking for any signs of the *Burnell.*"

He sounded so worn out and stressed that Ricci exchanged worried looks with Charani and Bastié.

Bastié asked, "How much longer do you think the two of you can take this, Tal?"

"The breaks are good, we've both been able to grab a meal bar and some water. That's really helped."

"That didn't answer her question," Ricci said.

There was a long silence while Jacoway was obviously trying to quantify their mental and physical states. "Three more hours, tops...I think." Tal sighed heavily. "Then we're gonna have to come out and take a real break. How's the stability of the field?"

Lateef answered him. "Our model predicts approximately eighteen hours before final collapse. You and Decker need to be out of the vicinity by then."

"Copy that. Jacoway out."

***

Tal's estimate of three hours had been a bit over-confident. Two hours later the *Cerxai* headed out of the Expanse for a long break. The two pilots were too worn out to bother boarding *Lovelace* and simply rested and napped aboard the *Cerxai*.

Over comms Ricci and Bastié had a quick word with Jacoway and Decker, then Clemente took her turn to ask for some medical data. Rita said, "You're both showing advanced signs of stress and fatigue. Are you sure you don't want to take longer than a three-hour break?"

"No, sir," Tal assured her. "We'll be good by then. Right, Decker?"

"Absolutely. The sooner we find the *Burnell*, the sooner I can sleep for about, oh...I'd say, a day and a half."

"That I gotta see," said Kennedy.

After the *Cerxai* went back in, the *Lovelace* bridge crew watched and waited anxiously, hoping for good news every time they checked in but the search had thus far proved fruitless. Five hours later the *Cerxai,* piloted by a deflated Decker, came out of the Expanse for the second time. After both pilots had rested for another three hours, Clemente insisted that the ship dock with *Lovelace* so she could personally examine them. She did so on the *Cerxai*, under the watchful eye of Ricci, Bastié, and Charani.

While he took his turn at being scanned, Tal said to Zache, "Your concept worked. You must be ecstatic."

"I'll be a lot happier when I watch the *Cerxai* pull into this shuttle bay for good." Matt remained silent but privately agreed wholeheartedly with that sentiment.

Looking at her scans for both pilots, Clemente pronounced, "This next trip better be the last one."

Wearily, Jacoway said, "That's about all we have time for. Knowing that, I'm going to make our deepest run into the turbulent part of the Expanse, yet. Okay by you, Decker?"

Naiche had been busy greeting a joyful Kayatennae but took time to say, "Last chance – let's make it count." As the senior officers disembarked, Decker smiled at Ricci and said, "See you in five hours or less, Captain."

"Less sounds good to me," Matt answered.

As they were heading back to the bridge, Clemente asked him, "Exactly how much rest have *you* gotten over the last eighteen hours?"

"Worry about your patients, Doctor, not me," Ricci answered.

"I am worrying about one of my patients," Rita insisted. "Don't make me relieve you of command. Take a nap – VICI will wake you when they find the *Burnell*."

Ricci headed for his quarters, not bothering to quibble with either far-fetched idea – his napping or them finding the *Burnell* at this point.

***

With just under two hours to go until they had to call it quits, Jacoway thought he caught the blip of a distress signal on the scanner but just as quickly it was gone. Then, there it was again. He studied the area sensor data carefully; there was some kind of spatial disturbance. Those weren't exactly uncommon in the Expanse but this one could be masking the distress signal. When his turn for piloting came, Tal headed for the area that might have been the source of the distress signal. As they got closer, the distress signal grew stronger, though still intermittent. During his next break, Tal was positive it was the *Burnell* and tried hailing them. To his immense joy, there was an answer from Lieutenant Cate Longo of the *Burnell.*

"I don't know who you are," Longo said, "But I've got some really bad news for you."

"This is Lieutenant Commander Talako Jacoway of the *Cerxai,* and we have really good news for you," Jacoway answered. "We're a UDC rescue ship."

Captain Beatrice Arturo came on the line. After identifying herself she said, "How do you propose getting yourself out of this mess, let alone us?"

"I'm sending that information over now." After hitting send on the data packet, he asked, "How is your ship and crew holding up, Captain?"

"The ship's in one piece and the crew...well...we'd just about resigned ourselves to living the rest of our lives out in the Expanse. The most we'd hoped for was that someone would eventually find the data we've been collecting."

"Now, you can hand it over yourselves." Jacoway knew his turn at the controls was coming soon and explained that they'd be talking to Lieutenant Decker next. "Did you get the data we sent?"

"No. But there's some kind of strong anomaly distorting the field near us. We were investigating when we got your hail. We'd almost forgotten that we're still broadcasting a distress signal."

Jacoway re-sent the tether instructions and took his turn at the controls. By the time his next break came, he got the happy news that the *Burnell* was headed out of the Expanse. He was able to relay the news to the *Lovelace* and breathe a massive sigh of relief. His elation didn't last long as heard Decker snarling, "What the fuck is your problem?"

He looked at her in surprise and annoyance that immediately turned to concern when he realized she was talking to the ship. "What's wrong?"

"The goddamn thing is fighting me all of a sudden!"

Before he could investigate, an alarm sounded from the control panel. "What is it?"

"Ship's not sure but something is seriously disturbing the Nav-sat," Decker answered.

"Yeah, you're right," he confirmed, looked at the data outputs. "Can you move away from it?"

"What a good idea. I wish I'd thought of it," Decker responded.

Tal didn't answer but wondered if there was any situation so serious that Decker actually refrained from resorting to sarcasm. He watched her hands flying over the panel and she finally sighed, "Phew, I managed to shift us to a better place. Looks like the Nav-sat is back in business but the sooner we get the hell out of this region, the better off we'll be."

With only minutes until he took the controls himself, Jacoway checked in with the *Burnell*. Longo answered, "We lost the tether there for a second, *Cerxai*. Scared the hell out of us. What happened?"

"Whatever is in the vicinity here was messing with our Nav-sat. We had to change position."

"If at all possible – could you *not* do that again?"

"We won't," Tal assured them, as he took the controls. However, he quickly regretted that promise since the ship almost immediately started fighting him as it had Decker. He had a dim awareness of her scanning all the data outputs from the ship, desperately trying to figure out what was going on. Most of his attention was focused on maintaining their present location and keeping the tether intact. His attention was jerked away from navigation when Decker announced, dread icing her voice, "Oh my God, it's a fucking magnetar. We're in danger of getting pulled into the wake of a magnetar!"

Chapter 18

Weep I Cannot

"Weep I cannot; But my heart bleeds...." William Shakespeare, The Winter's Tale

"That's impossible," Tal shouted. "A magnetar doesn't just appear out of nowhere!"

Naiche didn't know what to tell him. On the one hand he was right – if they were anywhere near a neutron star with a magnetic field that powerful, it should have been evident long before that moment. On the other hand, it was the only conclusion possible from the data she was seeing from the sensors. "Apparently in the Expanse it does." Decker hit the comm to *Burnell*. When she got Longo on the line she asked, "What's your ETA on getting out of the Expanse? We need to bug out of here as soon as possible."

"Copy that, *Cerxai*, we'll speed it up. We should clear the Expanse in thirty minutes or less." Decker thanked them and took her turn at the controls.

Twelve minutes later she flipped piloting back over to Tal, warning him, "The pull is stronger than ever; holding this position is really getting dicey. Do you think we have time to move, find the *Burnell,* re-tether them and still have everyone make it out of here?"

"We'll need at least an hour – preferably better than an hour to be safe. Check with *Lovelace.*"

To her dismay, Decker checked in with Lateef and learned that they now had forty-five minutes to get both ships out of the Expanse. "Okay that's not an option," she said. "The *Burnell* has to move its ass for real now. I'm going to contact them again."

She did so and heard back from Longo, "We're already at top speed. We were able to contact *Lovelace* and they tell us we've only got about another fifteen minutes to go before we cross over. Can you hold, *Cerxai?"*

Decker looked at Jacoway, knowing that he'd been listening to that exchange. He nodded yes. She relayed the affirmative to the *Burnell*, wondering if Jacoway thought he could hold position and stay safely away from the magnetar or if he knew, as she did, that with their two lives stacked against the fifty-five on the *Burnell*, they had no choice either way.

Eight minutes later they were in sight of the magnetar and no longer had a choice of moving or not moving. The *Cerxai* was caught in the powerful magnetic field and Decker knew that if they didn't find a way to fight free, the small ship would be scorched by the constant eruptions of gamma rays and X-rays from the magnetar.

They got the good news that *the Burnell* had made it out and Decker calculated that they didn't have much longer before the magnetic field caused them to lose communication with the ships outside of the Expanse. She turned her attention to piloting, knowing it had to be her turn again by now but to her surprise, control still hadn't flipped. "The switch isn't working."

"It's not working because I turned it off," Tal announced.

"What? You can't fight this damn thing all by yourself! Even if you manage it, you'll fry your brain to a crisp."

"We're not gonna get out of here switching back and forth. One of us has to take point on this. I'm not turning over control!"

"For fuck's sake – why does it have to be *you*?"

"Because I said so, Lieutenant!"

***

To his surprise Ricci had actually managed to doze a bit when VICI woke him, informing him that contact had been made by the *Burnell*, announcing that they were on their way out of the Expanse. He took the bridge in a jubilant mood, accompanied by every member of the command staff, all eagerly awaiting sight of the long-lost science vessel. Cheers rang out in every corner of the ship when, after Bea Arturo hailed the *Lovelace,* Ricci announced that their mission had been a success.

Matt, an eager Kayatennae by his side, leaned forward towards the view screen, straining his eyes for any sign of the *Cerxai.* There was an expectant hush over the bridge when several minutes later Decker hailed them. Ricci said, "Talk about cutting it close, Lieutenant Decker! You have us all on the edge—"

A deadly serious Decker broke in, "*Lovelace,* I'm sorry to report that we are...." Everyone exchanged worried looks as she finally concluded, "Our ship is in trouble, Captain. Big trouble."

A frisson of alarm ran down Ricci's spine. He fought to keep his voice calm, asking, "What kind of trouble?"

"We're caught in the wake of a magnetar. It does *not* look like...we're not going to be able to break free."

While the bridge broke out in exclamations of distress, Bastié leaned into the comm at Matt's chair, "Decker, have you tried—"

"Tal's tried everything, sir. You should all know that he's fought this thing all by himself and I've never seen better piloting in my life. If he couldn't do it – no one could. He's still fighting it or he'd...he'd tell you so himself."

"Why did you let it get this far?" Charani cried out, her voice rough with anguish.

"We had to – anything else would have broken the tether. We knew all along that this outcome was a...a strong possibility." Matt could hear his daughter's composure breaking down and tears welled in his eyes. Decker took an enormous gulp of air, and in a shaky voice assured them, "This was our choice. We knew what our options were and we chose to hold. I hope knowing that, helps all of you find peace with it." In a stronger voice she added, "Con, remember your promise."

"I will," Kennedy answered. He dashed an arm across his eyes before offering, "You sure made us all damn proud."

Decker answered, "*Yadalanh, shilsáásh.*"

Con choked out, "Goodbye, Deck," and then broke down crying, turning into his wife's arms.

Still hoping to awaken from this nightmare, Ricci protested, "Naiche, there's *got* to be *something*—"

"Just remember, Captain. This wasn't a repeat of the Tripoli mission – it worked. This mission was a success." Matt was standing stock still, speechless with horror when Naiche suddenly blurted out, "Love you, Pop."

"Love you, too," he answered automatically, as the comm line went silent.

For the first time in his life, Matteo Ricci froze in the line of duty. For an entire minute, he had no idea what to do or say. He stayed where he was, on the deadly quiet bridge, staring out at the view screen where the two rescued ships were still visible. Finally, Lindstrom came over to him and said, "Captain, if you need me to—"

"Yes, Commander, take the bridge." Matt stumbled away, dimly aware of Jackie moving to follow him before being waved off by Zache. He didn't even have the wherewithal to be grateful to her.

***

Decker watched a sweat-soaked Tal still desperately piloting against the pull of the magnetar, wondering if their death would be a quick one. She hoped so.

Suddenly, optimism breaking through a voice that was slow with fatigue and pain, Jacoway said, "Hey, that's...."

He didn't finish his thought but Decker looked at the data screen to see what, if anything, had changed for the better. "The field around us is shifting! Is it helping?"

Jacoway nodded, saying, "The shift moved us far enough away enough that I should be able to...." The ship started bucking, each jump getting them further from the magnetar. "It's working – we're getting free!" He rubbed a hand over his neural port, wincing in pain. "Decker, you need to take over," he begged. "I can't stand much more of this."

Deck went to flip the controls to herself but it didn't work. "What the hell...?"

"Decker, take over!" Tal screamed. "Now!"

"I'm trying," she shouted back. Her hands skated rapidly over the control panel, trying everything possible, but the stubborn Nav-sat remained fixed on Tal's port. "It's stuck – you're caught in some kind of feed-back loop with the QNS!"

It was clear Tal didn't have the bandwidth to respond to that but instead shouted navigational instructions to her while his right hand was clamped over his port. Decker obeyed, piloting the ship as he commanded but also watching helplessly as he writhed in pain. Following his directions, she managed to get the ship completely free of the magnetar. With a scream of agony Jacoway collapsed onto the control panel.

To Naiche's horror, the Nav-sat unit was still resonating back and forth with his neural link and when she reached over to feel the port, it was almost hot enough to burn. She knew time was short before Tal suffered permanent brain damage or even death. She tried cutting power to the QNS but there were too many fail-safes in place. All out of elegant technological options, Deck grabbed a wrench from the tool kit and bashed the QNS emitter until she cut the ship's connection to Jacoway. Even out cold, his body's immediate relief was evident. She pulled his unconscious form out of the seat and laid him gently on the deck.

Decker sat beside Tal, breathing heavily. The *Cerxai* was free of the magnetar but now they were deep inside the Expanse – with no QNS drive and no tether, leaving them no way to communicate with the ships outside. Naiche also knew that the field shift that had freed them signaled the start of the final collapse of the Expanse – soon to put them well out of the reach of any UDC help at all. *All right, Decker, you've been in tougher spots than this...what're you gonna do? Think.* The only thought that occurred to her was that no – she'd never been in a worse situation in her life.

***

After ten minutes alone in his office, Ricci had composed himself enough to return to duty. He went back to the somber bridge and took command again. He ignored the shame he felt at freezing up and tried to act like the captain he was supposed to be. Lateef informed him that the Expanse was shifting and would disappear from that sector of the galaxy in fifteen minutes or less. "We'll hold here until then," he ordered.

He then took an on-screen hail from Captain Arturo. After greeting Matt and accepting his congratulations on escaping the Expanse, she said, "Captain Ricci, I understand that your daughter was one of the two pilots who rescued us?"

"That's correct," he answered, fighting to keep his voice steady.

"You have my deepest sympathies for your loss. When we return to Uniterrae, I intend to nominate both Jacoway and Decker for the Cortez Medal of Valor."

Matt closed his eyes in pain for a second, seeking the strength to thank her. The Cortez medal was strictly awarded posthumously – to those who sacrificed their own lives to save others. "I appreciate that, Captain Arturo. They certainly deserve it."

"They sure do. Their bravery will never be forgotten by me or my crew. I want you to know that we would *never* have asked them to maintain position if we knew that...if we knew what it would cost them."

"I understand. I'm sure that's why they didn't let anyone know their situation until you were safe." Ricci took a deep breath and then informed her, "We will be setting course for Uniterrae as soon as the Expanse has disappeared and I have time to make a shift change."

"I've spoken with Captain Mazuzan, and if it's all the same to you — we think we'd like to take off immediately. Our apologies, but the *Meitner* is a much slower ship and we—"

"No apologies necessary, Captain. We'll see you back at The Rock."

The bridge crew watched both ships make the jump to L-speed and then the *Lovelace* was alone again. Kennedy looked at Ricci, asking, "Do you think there's a chance...?"

"I don't know. We'll give them this time and see." No one on the bridge made any pretense of doing anything but waiting. As the minutes ticked down, Ricci looked over at Kennedy, who was splitting his time between Aqila and Kay, since the sensitive dog had definitely picked up on the mood of the surrounding humans and had become visibly dejected. Matt was well aware that Con's suffering was as deep as his own. Ricci would have liked to reach out to his daughter's best friend but knew doing so would destroy his thin veneer of composure. Commiseration would have to wait – Matt couldn't risk losing his self-control on the bridge again.

Ten minutes later Lateef sadly announced that the Expanse collapse was complete and it had disappeared from their sensor range. Ricci swallowed once and got up from his chair. He straightened his uniform and said, in as assured a tone as he could manage, "Petrović, set course for Uniterrae and then turn your console over to beta shift. Everyone else, please do the same." He cleared his throat and turned to his first officer. "Lindstrom, you have command for the time being. I'm not to be disturbed for anything short of an emergency." He added firmly, "That's an order," defiantly eyeing his command staff. The last thing he did before leaving the bridge was to call Kayatennae over to him. Man and dog silently, slowly, walked to his quarters together.

Chapter 19

The Constantly Shifting Truth

"If you are terribly truthful, the ground will always move from under you, and you will have to shift with the constantly shifting truth." Anais Nin

Tal woke to the sound of...chanting? He focused and then recognized Decker's voice. "*Gą́hé, cheełkéń nts'ís dighį, Beenzhǫ́go 'ánágódlá, Gots'íse 'ádeenágódlá, Gogálei ká'ánágódlá.*" She was definitely chanting in what he assumed to be Chiricahua.

Having heard Apache prayers through his association with the One Nation Collective, the cadence was vaguely familiar to him – though he couldn't translate the words, of course. He opened his eyes and saw the gray overhead bulkhead of the *Cerxai*. Tal took a deep breath, noting the pounding in his head that accompanied his overall pain and weakness. He rolled his head to the side and was able to see that his uniform jacket had been removed, leaving him in his regulation t-shirt. There were two pain patches on his arm.

Suddenly, Decker's face loomed over him. "How do you feel?" she asked.

"What size was that missile? The one that hit me, I mean."

"Since you're well enough to joke, do you think you're well enough to sit up?"

"Give me a minute." Tal looked at her and asked, "Were you...praying? For me?"

"Umm...yeah." She studied him for a second before asking, "Do you find that offensive?"

"No, not at all." What he actually felt was a sudden but vague longing that he was familiar with his own people's prayers – but that wasn't a discussion he was up to having just then. Jacoway cautiously pulled himself into an upright position. Decker handed him a water tube and he slowly took a wary test sip. When it went down easily, he took a few more. His head gradually cleared and he asked, "Are we on autopilot?"

"No, we're parked on a small moon I found that can maintain human life. Well, not maintain it but at least support it."

"What's our status?"

"We're stuck in the Expanse with no QNS-drive. On the upside – we're both still alive."

"Wait a minute!" Tal exclaimed. He winced as that action had been unwise considering his mental and physical state. In a calmer, quieter tone of voice he asked, "What happened to our QNS-drive?"

"I bashed it in with a wrench."

It took Jacoway a moment to parse that statement and another one to accept that she wasn't kidding. "And you did that because...?"

"Because your port was stuck in a feedback loop with it and it was going to kill you."

Torn between appreciation and aggravation, Tal objected, "Not to sound ungrateful but you *do* realize that even if that had been true, which I don't think you could've known for sure – at least that way, *one* of us would have had a chance at getting out of here?"

"I was sure. Your port was a hot as a frying pan and as for your other point – no way! I don't know how it worked in Astra...whatever squad you were in, but in Force-1, we had a motto: 'Look out for the enemy but look out for each other even more'."

"We looked out for each other!" He grimaced at the pain that enthusiastic answer had engendered and scowled at Decker. "Is upsetting your patient this way also something you did in Force-1?"

"I wasn't a medic in Force-1." Not giving him time to respond to that news, she said, "You should eat something. You wanta try a meal bar or some bison jerky?"

"Is that a trick question? You really have bison jerky on you?"

"Yeah, I have it in my pack."

She rummaged through her daypack and handed a strip of jerky to him. He took a small bite. After a few more bites, he observed, "It's really good. Homemade?"

"Yeah, my cousin Josei gives it to me – her secret recipe." After eating several pieces of jerky and drinking an entire tube of water, Tal felt up to assessing their situation. "Why did you land on this moon?"

"I didn't want to wander around aimlessly until I can see about repairing the QNS-drive. So far that's been a bust. But still – it's like when you're lost in the woods. Don't make things worse by getting even more lost. Stay put until you've gotten your bearings."

"You know there's no way to get *your bearings* in the Expanse, right?"

"Look, it's not a perfect analogy, okay?" She finished off her own piece of jerky promising, "Don't worry, I'll think of something."

"No offense, but after that wrench thing, I'm not that impressed with your problem-solving skills."

"Too bad – I'm in charge now."

"You are?"

"Yeah, I relieved you of command due to your medical condition. Sorry – you missed that bit. Being unconscious and all."

Jacoway had no strong objections but still had to ask, "Does a medic *really* possess that kind of authority?"

"When we get back to Uniterrae – using my problem-solving skills – you can find out."

***

Ricci heard the chime at the door to his quarters and was tempted to ignore it. After all, if there was an emergency, surely Lindstrom would've informed him via VICI. Not wanting to take the chance, he answered the door to find Kennedy standing on the threshold. There was no way he was turning Con away so Matt invited him in and offered him a drink.

Con gratefully accepted a glass of whiskey and Matt decided to join him. The two men drank in silence for a minute before Kennedy asked, "How are you doing, sir?"

"I am...not sure to be honest. For the time being I think I'm running on denial and numbness."

"That makes two of us." Con nodded at Kayatennae who was stretched out motionless on the rug near the entrance to the bedroom. "Think he knows?"

"He knows something is wrong." Matt smiled sadly, explaining, "Mazuzan told me that he let Decker take Kayatennae with her when she left S and R because he wouldn't have been able to live with himself if he'd broken their bond." Matt took a sip, before saying, "Now fate has gone and done it." He looked at Con, asking, "Was that the promise you made to Naiche? The one she told you to remember – that you'd take care of Kay for her if...if she...." He couldn't force the words out. The best he could do was to finish, "If the worst happened?"

"No, sir, it was actually that I would look after...well, you," Con admitted sheepishly.

Matt had to take a few deep breaths to hold back the tears at that news. "I guess this visit is the start of that obligation?"

"I don't think of it as an obligation, Captain. I think of it as a privilege."

Taking another swallow of whiskey, Ricci whispered, "I guess she would know how badly I need looking after if she could have seen me on the bridge today."

"Sir?" Con asked, obviously puzzled.

"When I froze up. After she...after we lost contact."

"That was a pretty reasonable reaction in my eyes – and the eyes of everyone else on that bridge."

"Not reasonable for a captain in the UDC."

"Really? Because I don't think that reaction means you're unfit to be a captain – I think it means you're human."

In spite of his sorrow, Ricci had to smile at hearing his words echoed back to him when he needed them most. "That's good to hear – coming from a future captain, and all."

"That's very kind of you to say so, sir, but may be a bit of stretch, isn't it?"

"No, it's not. What do you think I saw in you when I recruited you? My eventual replacement, that's what."

Kennedy shook his head forcefully, disclaiming the proposed honor. "I don't know about that, now. I don't know if I'll have the heart to go on in this job – alone."

"You're not alone, Con. You have Aqila, you have me...." Ricci nodded towards the dog. "...and you have Kay."

The younger man responded with the trace of a smile. "You're right. That's pretty much what Aqila's been telling me. I need to work on remembering that." He took a drink before saying, "I thought I was here to help you out. Instead, you're helping me. Apparently, I stink at this consoling thing."

"Not at all. The only way we're gonna get through this is to support each other, you and I."

"Then tell me how to support you right now."

Matt thought for a second, and said, "Tell me about your happiest memory of her."

After considering the matter in silence, Kennedy offered the story of the first time Naiche had admitted how much Con meant to her. He then asked for Ricci's happiest memory. Matt had the answer ready at hand. "When I realized last year that calling me 'Pop' came to her just as naturally as 'Captain' had before." The two men exchanged a few more bittersweet reminisces before calling it a night. Matt escorted his visitor to the door and found another one waiting in the wings. He said good-night to Kennedy and told Zache to come on in.

"Are you sure you're up to a second visitor tonight?"

"Third. Clemente stopped in to check on me earlier."

"The question still stands."

"Yeah, I'm up to it."

He offered her a drink which she declined. "I just wanted to see how you're doing – make sure you're not still beating yourself up."

"Beating myself up?"

"For having that human moment on the bridge. I could tell that bothered you."

Matt sighed and rubbed his hand across his forehead. "Yeah, Kennedy and I already talked that out."

"Good. I hope he made you see that no one blamed you for needing a moment after exchanging your last 'I love yous' with your daughter."

"First."

"What?"

"First *and* last."

Zache pulled him into a gentle hug, whispering, "Oh, Matt."

He returned the hug, admitting, "You know...I never even thought to wonder – if she loved me. It was enough that she let me...love her."

"You're a hell of a man, Matteo Ricci." She pulled back from the hug and studied his face. "Do you want to talk some or do you want to get to bed?"

With a sigh, Matt said, "I suppose I should try to get some sleep. I *am* still captain of this ship and I have duties to get back to tomorrow."

"Message me if you can't sleep. We'll play cards or run the passageways or do whatever it is you need to do." Zache walked slowly to the door, and paused at the threshold, turning back to say, "Okay, I swore I wasn't gonna ask this – but I have to. Matt, do you...do you blame me at all?" Before he could answer, she said, in a rush, "I'd certainly understand if you do."

"No." Charani gave him a skeptical look. "Okay, maybe if you had asked that question on the bridge right afterwards, I would've had a different answer. But after thinking it over I realized Naiche said that part about it all being 'their choice' for a reason. So, I'd remember that this mission was something she took on of her own free will. Because that's who she is...." Wincing at the stab of pain that slip caused, he amended, "Who she was."

"There was more of her mother in her than many people realized," Zache said.

Matt held up the last of his whiskey, offering a toast, with just a trace of bitterness. "To the Decker legacy – women too brave for their own good."

Charani nodded, and added, with extreme tenderness, "And to the Ricci legacy – the man strong enough to love them both."

After she left, Matt stared at the closed door for a second before concluding, "Strong enough and lucky enough."

***

Under Jacoway's watchful eye, Deck feverishly worked on the QNS emitter. "I might be able to rig it up so it at least works with the transdermal link," she said.

"If you'd let me get over there and—"

"You gotta be kidding. It's a miracle your brain isn't so much cornmeal mush right now. No, you need to rest. Doctor's orders." Before he could voice his obvious objection, she added, "Medic's orders."

"Do you really think you can fix that thing?"

"Well, as my grandfather – who *was* a doctor, used to say, 'Where there's life, there's hope'." She smiled at him. "That goes for you, me, *and* the QNS." Deck returned to her work muttering, "I do regret not paying more attention in those engineering courses, though." After catching her finger on one of the circuit boards, she shook the pain off, announcing, "My biggest regret though is learning how to fly this damn thing in the first place."

"No, that's on me," Jacoway lamented. "You're here because I nominated you for this mission."

Decker slotted the repaired board back into place and faced him. "It is *not* on you. I took this mission on *voluntarily*."

"Yeah, I heard that speech you gave to the *Lovelace* – very moving."

"You caught that?" she asked, her attention already fixed on the next board needing repair.

"Yes, I did...." Amusement crept into his tone as he added, "And I heard you tell your father you loved him."

With a huff of irritation, she asked, "And that's funny to you, why?"

Tal made the sudden switch to sincerity, answering, "It was actually very sweet." She rolled her eyes at that descriptor and he ventured, slyly, "Though I never knew you called him, 'Pop'."

She shrugged, donning a nonchalance she didn't fully own. "Did you *really* think I called him 'Captain Ricci' all the time?"

"Honestly? I woulda guessed you called him...I don't know...something in Chiricahua."

"*Shitaa*? I do call him that when we're visiting my family."

"He goes to Chiricahua territory with you?" When Naiche confirmed that he often did, Tal asked, "Does he speak the language?"

"He's learning." Jacoway remained silent, so Decker asked, "Why?"

"I've decided that I want to learn, too. Not Chiricahua – Choctaw. If we don't get outta here, I guess missing that chance is gonna be *my* biggest regret."

Decker put down her tools and stared at him. "Really?"

"Yeah, after some stuff Sasaki said, and seeing how you enjoy the full richness of your own heritage...." He stared off into the distance for a second before looking back at her to firmly state, "Not that I think your band of Chiricahua's way is one-hundred percent right, either. There's gotta be a middle ground – somewhere." He seemed uncomfortable with the subject and changed it, saying, “Speaking of the richness of your heritage – is there any more of that jerky?”

“Sure. I’m ready for a break anyway.”

While they were eating Decker ventured, “You know...my mother thought there was a middle ground, too. Between being isolated and being assimilated, I mean. She believed she could be Chiricahua and still belong to the wider world.”

“Is that right?”

“Yeah, that’s why she joined the UDC in the first place.”

“I was always a great admirer of your mother. I don’t know if I ever told you that?” When Deck shook her head, he added, “I was actually kinda intimidated at finding her daughter in one of my classes.” Naiche almost choked on her jerky when he said, “For the longest time, I was trying to work up the nerve to ask you out.”

After staring at him to see if that had been some sort of bad joke, she exclaimed, “Then why did you turn me down when I asked you out?!”

“You *didn’t* ask me out,” Tal insisted, “you propositioned me.”

“Oh, I’m sorry, *your highness*. And the difference would be...?”

His dark brown eyes flashing with ire, he explained, “The difference is, when you ask someone out – to a vid, or a concert, or even just for a walk, you want to get to know them – as a person! It means that you’re interested in more than just mashing genitals with them!”

“I really don’t think that was the phrasing I used....” When Jacoway just glared at her, she tried to get off the defensive by saying, “You could have explained that to me, you know.” His scowl only deepened and the light suddenly dawned for her. “Ohh...that’s what you meant when you said you needed to get know someone first. You wanted to get to know *me*.”

Tal nodded sadly at her, observing, "That never even occurred to you, did it?"

"No." Naiche groaned, "Oh shit. I hurt you." When he nodded again, she said, "I'm sorry – I didn't mean to."

"I know...but it still hurt," he admitted. "*That's* why I was trying to trade study partners with Bian. It actually had nothing to do with your command of the Standish language."

Her head spinning with this complete realignment of past events, Deck breathed, "Wow. That's...." She didn't even have words for the mixture of sorrow, confusion, and discomfort she was feeling. Venting her frustration on the only culprit she could identify, Naiche fumed, "This is why I *hate* this romance stuff. When it's just sex, a good time is had by all – and *nobody* gets hurt!"

Tal put a hand on her arm, asking, "Decker, why did we take on this mission?"

Puzzled at the non sequitur, Deck looked at him with narrowed eyes. "Does this have something to do with the concept of hubris? Because my father—"

"No! We did it because it was worth the risk. Saving those people was worth the risk we took. And love is the same way."

Tucking all of that away for later contemplation, she said, "You know...comparing love to this mission? I'm not entirely sure you're making the point you think you're making." Jacoway laughed but offered no further defense of romance. Decker said, "Anyway, I gotta get back to work. So, you can have that chance – to learn Choctaw."

Chapter 20

Leap of Faith

It takes a leap of faith to get things going
It takes a leap of faith you gotta show some guts
It takes a leap of faith to get things going
In your heart you must trust – Bruce Springsteen, Leap of Faith

After several hours of intensive repair effort on the *Cerxai's* Nav-unit, Tal watched Decker testing out the revamped QNS-drive using the transdermal link. She had actually been forced to hard-wire the link into the navigation system to get effective communication. "Well?"

Naiche cocked her head, causing the wires that connected the band on her forehead to the control panel to jiggle merrily. "I think it's working – after a fashion."

"How comforting."

"It only has to work well enough for us to reach the margin of the Expanse and get the hell out of this death trap. Once we're in a normal space-time field, it should work just fine."

"And then what?"

"And then we can actually get our bearings. Figure out where we are."

"Yeah – and possibly find ourselves fifty-thousand light-years from home."

"If that's the case, then we'll just have to locate a nice planet and start the human race all over again." She stripped off the transdermal band and turned around, assuring him, "That was just a joke, by the way – I wasn't propositioning you."

"Don't worry – my pearls are completely unclutched."

"Okay...whatever that means. I'll just assume it's a good thing." Decker stretched and yawned. "I'm gonna get some sleep so I can start out fresh." When Jacoway yawned back, she said, "You need to do the same – your body is still trying to heal."

Tal would have liked to protest but he was exhausted even though he'd pretty much only watched Naiche work on the QNS, while offering suggestions. He snuggled down on his makeshift bed next to the one Decker was setting up for herself and fell into a restless sleep. It didn't seem like very long before he heard Naiche moving around. He looked up to find her expertly redoing her braids in the semi-darkness. "How long have we been sleeping?"

While fastening her hair on the top of her head, Deck answered, "About four hours. You don't have to get up."

"No, I'm good," he claimed, though his limbs still felt fairly leaden. Jacoway and Decker breakfasted on the last of her jerky and he noticed her swallowing a white tab that she'd taken out of a small packet. Before putting it away, she was carefully counting the remaining tabs.

"What are those?"

"Caffeine tabs – want one?"

"No, thanks." She seemed relieved at his refusal so he suggested, "If you need those to stay awake them maybe you should get some more sleep."

"I need them to ward off a migraine."

After considering that news for a second, Tal proposed, "You're probably just addicted."

She shrugged, admitting, "Probably, but life without caffeine – is that really living?"

"What happens if this planet we might need to settle down on doesn't have caffeine?"

"I said a *nice planet.*" With a laugh she asked, "What kind of self-respecting planet doesn't have caffeine?"

A little while later they were taking off, and Tal held his breath until Naiche announced that the jury-rigged navigation system was working. "I'm going to have to take it slow though – that's the only way it's gonna work with the transdermal link."

Jacoway kept an eye on the sensors and data out-puts while Decker shouldered the monotonous task of slowly piloting the *Cerxai* through the Expanse. When they broke for lunch, he asked about taking his turn at the controls.

"It's best you continue to rest and heal. Your last med-scan still showed some synaptic damage and neurotransmitter imbalances."

In his weakened state, Jacoway couldn't really argue with that but he felt guilty about her bearing the entire burden. When he presented that sentiment to Decker, she assured him that piloting with the transdermal and at the slower speeds was nowhere near as taxing as the direct neural link had been. "I'm much more worried about our fuel situation," she said.

He put his meal bar down, asking, "Why? How bad is it? We couldn't have used that much of—"

"We burned through over half of our reserves fighting that magnetar."

"Shit! What's our rate of consumption now?"

"Well, the only upside to moving this slowly is that it's pretty low."

"But the downside to moving slowly is that it's going to take us that much longer to get out of the Expanse. And when we do, we'll still have to get...at least, *near* a relay station, to have any hope of rescue."

"I know." She gave him a somewhat forced smile, saying, "Look, we'll burn that bridge when we come to it. First things first – get the hell out of here." She nodded at his hand. "Finish your meal bar. You need that nutrition."

Reluctantly, Tal picked it up and resumed his desultorily chewing. "Ugh. What do they put in these things?"

"Only the finest sawdust," quipped Decker, as she polished off her own bar.

"I'd kill for a piece of my dad's Navajo frybread," Tal said. He looked at her, stating defiantly, "By the way, I plan to embrace my Choctaw heritage but I'm *not* giving that up."

"Oh, we *all* make that!" Naiche exclaimed. As she performed a minor course correction, she informed him, "My Great-Aunt Loza makes the best frybread in the world."

Jacoway said, "I guess we'll just have to see about that."

Deck raised an eyebrow at him, asking, "You angling for an invite to Chiricahua territory?"

"Maybe."

To his surprise, she answered with evident sincerity, "Sure thing – right after we get home."

The mention of home prompted Tal to perform some precise calculations regarding their course to the nearest margin of the Expanse. After viewing the results, he announced, "At our present speed, and with the current configuration of the Expanse, it will take us fifty-two-point-five hours to get out of here."

"That's pretty close to my initial estimate. But I can't risk going any faster, otherwise I won't be able to keep up with the Nav-sat. The important thing is, we'll still be okay on fuel, correct?"

Jacoway checked the fuel reserves against their current rate of consumption. Decker was right – they would have enough to get out – it was their fate afterwards that still concerned him the most. "Yes, but, we better come out of the Expanse really close to a relay station – or a caffeinated planet to settle down on."

Deck was silent for a moment, but finally looked over at him and bravely quipped, "Got it – plotting a course for Caffeine World, well known for its lovely frybread fields."

"You always laugh in the face of danger?"

With an unusually serious edge to her voice, she admitted, "Yeah...it helps keeps the fear at bay."

Touched that she had allowed him that glimpse of her vulnerability, Tal conceded, "I know. I've whistled in the dark many times, myself." They exchanged brief smiles, acknowledging their shared experience.

Since there was nothing much to see or do while Decker was slowly piloting them through a region of relative stability, Jacoway found himself dozing off. He woke with a start when he heard Naiche saying, "Okay, look, it's nothing personal, I'm just not that kind of girl." Jacoway stared in amazement at the sight of her covered with the pink Jayhine dust.

***

"Decker, they might be able to help us!"

She hated to admit it but Jacoway probably had a point. The alien dust entities had never shown any malicious intentions towards humanity and they certainly knew their way around the Expanse. "Can't they help without invading my mind?"

"It's not an invasion. It's simply the way they communicate."

"All right, all right." She whispered one of her father's favorite phrase – "Fuck my life," and then said, "Okay, folks, welcome to the jungle. And *do not* make yourselves at home."

Decker immediately felt a question form in her mind as to why she "hated" the Jayhine. "I do not *hate* you, it's just that I had a bad experience with...umm, you know, the mind melding types."

Since Decker was thinking of the Pakarahova melds they had access to those memories. "Yes, I realize it doesn't seem that bad – it has a very bad association for me, though." Since she had absolutely no desire to "discuss" the comrade kills with *anyone*, let alone dust aliens, with great effort she forced her mind onto their current predicament.

Her head then buzzed with questions about how they had gotten trapped in the Expanse. Decker quickly ran through the events of the rescue and the magnetar. "No, we didn't know those people – not *personally* anyway, but we...that's what we do. We help one another – if we can."

The Jayhine wanted to know how Tal and Naiche *knew* that the humans on the *Burnell* and *Meitner* wanted to be rescued that badly. "We...um, extrapolate from how we'd feel in a similar situation." She listened to their objections and then said, "No, it's not a fool-proof system but we've gotten pretty good as a species at figuring each other out...." They had more quibbles to make so she said, "Look, when we can't do that...then...then we just give each other the benefit of the doubt."

Decker cut the philosophical discussion short by asking clearly and directly, "Can you help us exit this quantum entanglement – *quickly*?" She sensed their affirmative response. "Yes, they can," she narrated for Tal. There was an immediate inquiry from the aliens about the direction. "It absolutely matters where we end up when we leave here. Why? Do you—" Deck paused and listened to what the

Jayhine were telling her. "Of course, you do." She told Tal, "They say they know a short-cut – not just out of here but back to our sector of the galaxy. Damn, these guys are all about the short-cuts – aren't they?"

"Well, their last one was golden, wasn't it?"

"Good point. Okay, we're game. Lead on, folks."

The Jayhine eventually settled down to a low buzz in the back of her mind and Decker pretty much ignored them when they weren't actively giving her directions. At one point she looked over at Tal and saw that he was dozing beside her. The sight made her smile and she thought about the many twists and turns of their history together. Suddenly the Jayhine were afire with questions about her and Jacoway.

With a groan, Decker cursed herself for the mental slip but did her best to answer the questions the aliens had. *Yeah, you're right – I didn't always give him the benefit of the doubt. Well, I didn't know him that well. It's different with people you know, where there's love and trust. With them we can bridge the distance between us by assuming they have a good reason for what they do – even when we don't understand it.*

The question that formed next was startling and Decker never knew whether it was her own or not – *"Then, why don't you do that?"* Her hands actually slipped off the controls in shock. She looked guiltily at Jacoway but he was still sleeping and hadn't witnessed her agitation. The Jayhine, though, sure had. Her mind hummed with questions about the situation between her mother and father. *Shimáá* had her own reasons for her conduct regarding Ricci, reasons that Naiche would never fully know – but based on the loving, generous woman she'd been, they'd most likely been benevolent ones. Why couldn't she accept that?

And Cat Xavier, Ato Gbeho, and the others she'd been forced to kill. She'd never know for sure what those corrupted Pakarahova had whispered in their minds to convince them to accept the bonds – would she? Why not assume those people she'd fought side-by-side with, whose honor and courage she'd frequently witnessed, had their reasons, too? As she pondered this new reality, Naiche felt her anger at all her dead friends slipping away, being replaced by an aching acceptance. She wiped away tears streaked with pink dust and managed a shaky laugh. "I did warn you guys that it was a jungle in there."

At the sound of her statement, spoken aloud, Jacoway woke with a start. "Where are we? How close are we to reaching the margin?"

"Good question," Deck answered, wrenching her attention back onto their journey. She relayed the query to the Jayhine. "They say we're almost at the portal," she reported.

"The portal – what is that?"

"I don't know. I guess we'll find out when— Holy fucking shit!" Deck exclaimed, as she came face-to-face with a huge cosmic pool. "You want us to go into that thing?!"

***

"No, no, no, tell them, no, Decker!"

"I DID! They're *insistent* that this is the short-cut back to our region of space."

"Did you tell them we won't survive going in—"

"Yes! They claim we will." Decker waved her hands in a stop motion, saying, "Okay, folks, could you *please* quiet down and let me discuss this with my partner here."

Jacoway watched her sit silent for a moment, obviously listening to the Jayhine and then the dust aliens slowly glided off of her and coalesced into a ball behind her seat.

Naiche then looked at Tal. "Okay, what's your call, Commander? Because the Jayhine are hopping off at this so-called portal. They say this is the only way home for us."

"Are you...are you saying you want to try this crazy stunt?"

Decker sighed and stripped off the transdermal band and faced him directly. "When my grandfather died, he was only about seventy – still relatively young." Jacoway wanted to impatiently ask her what the hell this had to do with anything but instead let her talk. "I was devastated...I was seventeen at the time, and he had raised me and was my whole world after my mother died. When he told me his diagnosis, I was raging at how unfair it all was but he finally said something that helped me find some measure of peace with his impending death."

"What was that?"

"He said, 'I'd rather have a good life than a long life.'" She gave Tal a spirited smile, asserting, "I feel the same way. I think this might be our only way back home and therefore – it's worth the risk." Her smile slid into a smirk as she suggested, "Or maybe *you* can think of it the way you do about love."

"Wow, you really *do* have a low opinion of love," Jacoway muttered. He considered the quandary for a few moments but decided that Decker was probably right. If this didn't work – at least it would be a quick death, as opposed to them slowly dying, either trapped forever in the Expanse or stranded far from home. "Okay, let's do it."

"Really?"

"Yep."

"Okay, we need to ensure that all loose equipment is secured and then we better strap in tight – this is sure to be a wild ride." While they were working, the Jayhine disappeared from the ship and to Tal's amazement, Decker actually waved good-bye and thanked them. She apparently had a lot more faith in this portal idea than he did.

They took their seats and looked at each other. Jacoway nodded and said, “Okay, Lieutenant – punch it.”

Decker responded, “Talk to you on the other side, Commander.” She then donned the link and without hesitation, steered the ship directly into the cosmic whirlpool.

Calling the short-cut ‘a wild ride’ turned out to have been a major understatement. The *Cerxai* was tossed and turned and nearly shaken apart after entering the whirlpool. Tal could see Decker frantically trying to keep them steady but a few minutes in, the ship rolled and they were heading through the whirlpool sideways; it looked like Decker might lose control of the ship entirely.

Naiche shouted, “Son of a bitch!” while her hands flew over the piloting console, desperately trying to regain command of the ship’s direction. When she finally managed it, Tal didn’t have long to enjoy the respite as that same scenario was then repeated several more times. Jacoway started to think that they *had* chosen “good life” over “long life” when they decided to embark on this insane course of action.

His anxiety peaked upon detecting an ominous creaking sound from the inner hull; a second later an alarm sounded on the control panel. “Oh, just fucking great,” he groaned. Jacoway looked at the ship’s data output and bellowed over the din, “We’re at forty-eight percent hull integrity!” Decker made no sign of having heard him other than a nod, as all of her attention was riveted on piloting the wayward ship.

Silently, Tal watched as the hull integrity kept steadily dropping. When it hit thirty-three percent, he knew they didn’t have long before the ship gave way. As it fell to twenty-five percent, he braced himself for the inevitable collapse. Tal wished he had the opportunity to say good-bye – and more than that – to Decker. He was preparing to shout his farewell over the noise when the *Cerxai* shot out of the whirlpool and flew into what seemed like a stable area of space.

He and Naiche stared at each other, both gasping for breath and dizzy from the trip. He could do nothing but shake his head in astonishment and laugh when she said, "That was kind of fun – wasn't it?"

Chapter 21

Along the Common Ground

The oak and weed together rise,
Along the common ground
The mare and stallion light and dark
Have thunder in their sound – Richard Fariña, The Quiet Joys of Brotherhood

The Jayhine had once again proved themselves masters of cosmic short-cuts. The *Cerxai* was within striking distance of several Uniterraen outposts. Or they would have been, had their fuel situation not been so dire. Decker and Jacoway sat, studying the map display and discussing their options. Jacoway pointed at a relay station near Luhman-16, proposing, "We *might* make it to that one, if there's some fuel reserves beyond what's indicated."

"How accurate are the QNS-beta simulators?"

"Pretty damn accurate. Why?"

"Then we probably won't have enough fuel to both get there and to maintain even minimal support systems for more than a few hours."

"You got a better idea?"

Decker studied the map for a second, and then exclaimed, "Hey! If *Lovelace* took the predictable route home from where we all were last – and why wouldn't they? Then we should be able to cut them off near HIP-34074. They'll pick up our distress signal if we get out ahead of them and wait right there," she said, pointing to an area near the proposed star.

"Yeah, you're right; that *could* work." He tilted his head, obviously considering the idea carefully. "We'd have to time it perfectly, though. And if they took a different route—"

"Then we're screwed." She looked at Jacoway, awaiting his decision. When he didn't say anything, her impatience got the better of her and she exclaimed, "Well?!"

"What? Is this completely up to me?"

"Did I pick up a third stripe when I wasn't looking?" Naiche held her uniform sleeve up for inspection. "Nope, still just two."

"Come on – when did that ever stop you from voicing an opinion?"

"I did give my opinion. I like the *Lovelace* plan."

Tal laughed and said, "Okay, me too, but I think you like that one best because you get a chance to see your dad that much sooner."

Naiche was tempted to deny his supposition but there was little to lose with a plan this risky anyway so she said, "My dad, and my friends, *and* my dog." She looked at him, awaiting the inevitable ribbing. "*Happy*?"

Rolling his eyes, Jacoway just groused, "Oh yeah, if it weren't for my injuries, I'd be dancing for joy. Facing near-certain death is well worth it just to get you to admit the obvious." He took a deep breath and said, "Lieutenant Decker, set course for HIP-34074."

"We better perform the necessary calculations first, don't you think?"

"Yeah, I thought that went without saying."

They made it to their designated wait spot in eight hours and Decker put the ship in a holding pattern. She set the distress signal to begin broadcasting, quipping, "I hope it doesn't attract a Dardanze." After rechecking her hand-held, Naiche announced, "If our calculations are correct, *Lovelace* should be here within the next ten hours. Which gives us...." She checked their fuel status before reporting, "...about ninety minutes of life support as a cushion."

After almost two hours of waiting, the exhausted officers fell asleep, both hoping they'd awaken to a hail from *Lovelace.* Five hours later they were up and restless but working to maintain a hopeful outlook.

"I have a chess module on my hand-held; want to play?" Tal offered. Reluctantly, Naiche agreed, figuring it was better than checking the fuel status for the millionth time. A few moves into the game he noted, "You're not *that* bad for someone who hates chess. Why did you bother learning?"

"It was Lindstrom's idea – he was teaching me a lesson about being a wise-ass."

"I see." Jacoway laughed and said, "But the only thing you really learned from that lesson was chess, huh?"

"I can't argue with that." She watched as he gleefully captured one of her knights. "Why do you like it so much?"

"I like the strategic nature of the game and the fact that no matter how good you get, there's always more to learn...but mostly I like it because my mom taught me. It's still something we do together – long-distance most of the time." Since she knew nothing about them, Naiche then encouraged Tal to talk about his family, learning that his mother ran the Uniterrae Ministry of Health, his father was a governmental finance director, and he had an older brother who was also in Public Health.

"Wow, you're kinda the odd man out, aren't you?"

"We don't all take after our parents as much as you do, Decker."

"Do I? Cause I'm always hearing about how much I don't."

With a warm smile, Tal said, "I guess you're always hearing wrong then."

After two games, both of which Jacoway won handily, they switched to poker – revealing Decker's hidden forte.

"This is an *N'daa* game, too, you know," Jacoway reminded her, frowning down at his hand-held. "Why're you so good at it?"

"At the front, there were only so many ways to kill time between battles." Instead of automatically pointing out their very different war experiences, Decker dealt the next round of cards from the module on her hand-held, asking, "Didn't you fighter pilots play poker between missions?"

Without assuming a defensive posture, Tal admitted that fighter pilots always had access to rec centers. He then asked Naiche to describe the challenges of fighting on the ground. She did so – minus any potshots at his brand of service – and they whiled away a couple of hours comparing and contrasting their experiences. Before long, Naiche found the common thread in their stories – the war had left deep scars on both of them, scars no eye could see.

Eventually they could no longer ignore the fact that their situation was looking bleak. Decker had the ship's power completely devoted to maintaining position, broadcasting the distress signal, and minimal life-support. Even so, they had just under an hour before the fuel ran out. She reported the results of her findings to Jacoway and said, "I guess either we miscalculated or *Lovelace* took a different course home."

"Sure looks that way."

They sat for a few minutes in dejected silence, contemplating their approaching deaths. "Damn it!"

Jacoway exclaimed, while slamming his fist down on the control panel. "We came so fucking close to pulling this off."

"I know." Sighing heavily, Decker said, "It makes it hurt that much worse, doesn't it?" When Jacoway nodded in agreement, she bit her lip and then added, "I've faced death so many times, you'd think it would have gotten easier."

"It hasn't?"

"No," Naiche admitted. "This is actually the hardest one of all." She locked eyes with him and said, "It's like I have...more regrets this time around."

"Me, too." He took a deep breath and then said, "I feel...cheated, somehow."

"Yeah. There are important things...I'm leaving...unexplored." Deck was fairly certain they were both talking about the same regret but was afraid of saying more on the subject, afraid she'd break down. Which was definitely not how she wanted to spend her last hour. To change the subject – and the mood – she said, "I'm sorry you're not gonna get that chance to learn Choctaw." Tal threw up his hand in a gesture of glum acceptance. "At least you've always known one Choctaw word, right?"

"I have?"

Naiche sat up in surprise. "Yes...your name? Talako? It means 'gray eagle' in Choctaw."

"Oh right! Yeah, I think my grandmother told me that when I was a little boy. I never really thought about it again," Jacoway mused. "I was named after my uncle." He then turned to her, scratching the back of his neck, asking, "How do *you* know that? Is it the same word in Chiricahua?"

"No, not really. I know what your name means because...." After a pause, she sheepishly admitted, "I looked it up. You know, back when I had thoughts of seducing you."

"Oh." Jacoway nodded thoughtfully at that response. "I guess with time so short, I can finally admit that my *actual* biggest regret in life is that I...passed up my chance...." Tal hesitated while offering a wistful smile. "...to mash genitals with you."

After only a moment of consideration, Deck slyly ventured, "Well...that's one regret we still have time to fix." He stared at her, obviously uncertain of her meaning, so she explained, "I always wanted to go out with a bang."

With a sudden but eager grin, Tal said, "I hope you're not toying with me."

"Not at all – as long as you promise *never* to use the phrase 'mash genitals' ever again."

Holding his hand up, palm out, he swore, "Never. Not as long as I live. However short that may be."

"Of course...we will be burning through our oxygen reserves a little faster if we do this."

"Who cares?" Jacoway declared. "I'd rather live a good life than a long life."

"Okay, then." Deck rubbed her hands together and surveyed the cockpit. "Wonder how far back these seats recline?"

***

Captain Ricci was having dinner in the mess hall with Bastié and Charani, both of whom had joined him for breakfast and lunch, as well. He was torn between irritation and appreciation at the way his two old friends had been hovering over him on the journey back to Uniterrae. The nearest VICI unit squawked out, "Captain Ricci, your presence is required on the bridge. Priority situation." Matt muttered, "What now?" and threw his fork down.

Zache and Jackie chose to accompany him to the bridge. On the way, Zache asked, "With you, why does VICI always say, 'your presence is required somewhere' rather than 'report to that place'?"

"Because," Bastié explained, "Matt is the highest authority on this ship and therefore, no one can *order* him to report anywhere – but rather simply request it."

"Semantics," Matt groused. "My dinner is still interrupted." When they got to the bridge, he looked at Lieutenant Mikkelsen and asked, "Okay, what is so all fired up important?"

"We've detected a distress signal, Captain."

Snapping into action, Ricci slid into the command chair. "From a UDC vessel?"

"Yes, sir." Mikkelsen turned and faced him. "It's from the *Cerxai*."

Matt had to grab the arms of his chair to keep upright. "Are you certain of that, Lieutenant?"

"Captain, we've checked the U-dec signature five times. I am positive."

"Did you hail them?"

"We're not in hailing range yet, sir."

Swallowing to wet his mouth that had gone completely dry, Ricci hit the comm button on his chair and ordered the entire command staff to the bridge.

A short while later, Lateef was reporting, "We'll be in hailing range of the *Cerxai* as soon as we drop out of L-speed in...two minutes. Long-range scans indicate that the ship is...either out of power or running extremely low on power."

Ricci exchanged an anxious look with Kennedy. He was almost afraid to hope but he couldn't help it. He was praying for a miracle right then. If they came upon the *Cerxai* too late – Matt was pretty sure he'd never survive knowing how close they came to enacting a rescue.

When Petrović announced that they had dropped out of L-speed, he ordered Evans to hail the *Cerxai*, which wasn't yet in visual range. He waited, nerves on edge, and almost screamed in frustration when Evans reported that there was no response. "Lateef, can you scan for life-signs at this distance?"

"One more minute, sir."

"Evans, keep hailing them."

Evans nodded and repeated, "*Cerxai*, come in, this is the *Lovelace*. *Cerxai*, do you copy?"

Ricci gasped aloud with relief when the answer came back, "We copy, *Lovelace*, we copy. Commander Jacoway here." He sounded out of breath.

"Jacoway! We thought we might be too late there for a minute. Are you both okay?"

"Yes, sir, we're both fine," he declared. In a much quieter tone, Tal added, "If anything, you're here just *a little* too soon."

"What's that?"

"Nothing. Just a small joke, Captain."

"Oh." Ricci managed a relieved chuckle, venturing, "Lieutenant Decker must be rubbing off on you, Commander."

"What!? No, sir!" While Ricci was exchanging a puzzled look with Lindstrom, he thought he heard Jacoway say, "Ow!"

Lindstrom whispered, "There might be some oxygen deprivation at work here. Better have Rita check them both over thoroughly."

Before Ricci could respond, Decker was heard on the comm line. "Glad you're here, Captain. You guys showed up in the nick of time." His relief at hearing his daughter's voice erased every other consideration from Ricci's mind. "We're just about out of power and might even need some help docking. Will you be able to pull us in with the Bessel hook?"

Ramsey was on the bridge and said that Engineering would be proud to help. After a rushed, somewhat incoherent explanation about the Jayhine and cosmic whirlpools from Jacoway, the command staff left the bridge en masse, eager to greet the returning pilots.

When a disheveled, exhausted-looking Naiche stumbled out of the *Cerxai*, Matt pulled her into a bear hug and wondered if he could ever stand to let her out of his sight again. After half a minute, with Kayatennae frantically trying to break in between them the whole time, someone loudly cleared their throat. "I know this is an inappropriate greeting for a captain and lieutenant on duty but I don't really give a *damn*, right now."

"Actually, Captain," Kennedy responded, "I was wondering when you were going to give someone else a chance."

***

"No way! You are *making this up*," Con declared.

Naiche shook her head at him from her cross-legged perch on the Med-bay bed. "No, I'm not. That's why we took so long to answer the hails."

Kennedy stared at her for a second, apparently waiting for her to admit the joke but when she just stared back, nodding at him, he finally said, "Oh my God, you're telling the truth." He closed his eyes briefly, opening them to say, "I swear, Deck, that could *only* happen to you." While she laughed in agreement, he asked, "Is this gonna go in the report?"

"No, we decided to make that a discreet omission."

"Good idea." While Decker was petting Kay, who had insisted on getting up on the bed with her, Con asked, "So...did you guys resume...."

"No! The mood was killed for good. There was no *resuming*."

"Resuming what?" Deck and Con both looked towards the door to find that Captain Ricci had entered the room.

"Um, Deck wants to know when she can resume her duties, sir," Con claimed.

"We'll talk about that when you're released from Medbay."

Decker stretched and checked the chrono. "Any minute now, I hope."

"No, Clemente wants you here over-night for observation."

When Naiche groaned in response, Kennedy said, "I think that's a very prudent move." He looked at Ricci and said, in a mischievous tone, "She's been through *so much*. More than she can ever bear to tell you."

Decker was contemplating whether or not she could get away with smacking her CO in front of her captain when Con wished them both good-night and left.

"I really *have* to spend the night here?"

Ricci said, "Yes." He took a deep breath, shaking his head. "When I think of what almost happened...."

"That's kinda our fault. We should've remembered that you guys were moving slower because of the fuel burn from that Jayhine short-cut."

"Sounds like your short-cut was even riskier."

"Who knew those white-water rafting trips my grandfather took me on would come in handy in the UDC?"

"I'm convinced that the Jayhine don't really understand what is and isn't safe for human life."

"Yeah...probably not," Deck agreed. However, thinking of all she had come to realize in her time with them, she said, "But those little pink dust bunnies are *okay* by me."

"I know I'm forever grateful to them." He clasped her shoulder firmly. "They gave me my daughter back."

She put her hand over his, saying, "I'm sorry for what you've been through these past couple of days."

"No, don't be sorry. It's not your fault." Ricci straightened up, saying, "After we said...." He cleared his throat and started again, "After we said good-bye, I came to realize, it's no one's fault. It's the nature of this job – the one we *both* signed up for."

"Yeah, you're right." After he wished her good-night and was turning to leave, Naiche asked, "Why is that again? That we do this job?"

Matt paused at the door to say, "Oh, I thought you knew – we're crazy. It runs in the family." He smiled broadly, adding with a wink, "Just be thankful you didn't get my nose, too."

***

The next morning, a grateful Decker was released from Med-bay and cleared for duty. She stopped by Jacoway's room to check on him; she chimed at the door and upon hearing an invitation to enter, she ordered Kay to stay outside and walked in. She found Tal sitting up in the chair, entering data on his handheld. He reported that there were still a few tests the medical team wanted to run on his neural port before he could be released. Jacoway stood up and moved closer to her. His face graced with a shy smile, he asked about her condition.

"I'm good to go." She pointed at her own port, declaring, "I haven't decided whether or not I'm keeping mine. Makes a hell of a souvenir, though, doesn't it?" Naiche nodded at his hand-held which showed a full page of text. "Working on your report? Shouldn't you be resting?"

"Yeah, don't rat me out to Doctor Clemente but I wanted to get it down while it was all still fresh."

"Your secret is safe with me, but I hope you're gonna let me read it over before you submit it. I wanta check it for...umm, inaccuracies."

With a chuckle, Jacoway retorted, “I’m asking you to write your own report anyway, and then we’ll merge.” He added in a rush, “The reports, I mean!”

Failing to hide her amusement at his blunder, Naiche assured him very deliberately, “Oh, of course. We’ll merge...the reports.”

In an obvious effort to hide his embarrassment, he said, “By the way, I asked Clemente – UDC medics do *not* have the authority to relieve someone of command.”

“I’ll have to remember that. For the next time I’m stuck in a quantum entanglement – with a dashing lieutenant commander.” After smiling at his renewed blush, Decker kicked at the floor and ventured, haltingly, “So...when we get back to Uniterrae....”

“Yes?”

“There’s this vid playing at The Rock’s holo-theater about Maggie Heafield – the woman who designed the FTL drives we use. I hear it’s really good.”

Clearly mystified at the apparent change of subject, he just said, “Uh-huh.”

“So...I thought...maybe...you’d like to accompany me to one of the showings.” She looked at him hopefully, adding, “And then we could have some dinner, afterwards. My treat.”

Jacoway stared at her, a wide grin slowly forming on his face. “Lieutenant Decker! Are you asking me out on a *date*?”

“Yeah – why? Did I miss a step or something?”

Tal laughed. “No, it was perfect. I just think we passed the ‘getting to know you’ stage of things – don’t you?” He leaned forward, taking her hand and saying softly, “Naiche, I think it’s clear that we’ve already closed the gap between us.”

She smiled down at their entwined fingers, admitting, “You’re right. But we’re still gonna have that date. I wanta do things your way...at least once. Even if it’s not in the right order." She then gave him the kiss he had wordlessly requested. Their embrace was both tender and passionate; afterwards, they stood together, foreheads touching, savoring the moment. Naiche whispered, "I gotta go," and reluctantly left, calling back, "Let me know when you're free for our date and I'll arrange the entire evening." Deck stuck her head back in the room to add, "Or you know, the first part of it anyway."

Epilogue

The Journey We Make Together

We were never perfect.
Yet, the journey we make together is perfect on this earth who was
once a star and made the same mistakes as humans. – Joy Harjo, A Map to the Next World

Naiche pressed her palm against the lighted panel in the doorjamb of Ricci's quarters, wondering if her access would be revoked once she got her own quarters. When she was admitted, Deck found her father in the living room, reading. She parked herself across from him and greeted Kay. She then, rather unnecessarily, announced to Ricci, "I'm back!"

"Yes, I see. You got the present?"

"Yep. I had the receipt sent to your hand-held and the gift sent on to their quarters."

Ricci looked down at his hand-held and swiped to the receipt from the sporting goods store. "Holy shit! Is this right? *What* did you buy?!"

"It was on their list. Those sticks you use to play that *N'daa* game."

He cocked his head at her, obviously confounded. "You're telling me that baseball bats cost several *thousand* unnos?"

"Not baseball...um, golf! That's what it's called."

"Oh," Matt said. "You got them the matching custom golf clubs. Yep, that would explain it...."

"Yeah, I know it was a lot, but after all – you're their captain and it *is* from both of us." She laughed, "Well, sorta."

Staring at the receipt with narrowed eyes, Ricci said, "My, aren't *we* generous?"

"Here, this should soften the blow." Naiche pulled a bottle of *riserva* grappa out of her bag and plunked it on the coffee table. "I replaced your grappa. Finally."

"Yes, getting my grappa back – that *does* make up for spending a month's pay on a wedding present."

"Well, I could kick in a little...I guess." She got up and went to the kitchen for a drink. She came back into the room with an iced coffee, proposing, "How about you let me stay here for another month or two and I'll give you what I save in quarters-fees."

"Keep your money – a deal's a deal. And you can stay here for another year as far as I'm concerned. Or even longer."

"No, that's okay. I'm gonna get outta your way soon. I already started looking at available quarters."

"You're not in my way! I love having you living here."

"Then why did you ask me to leave?"

"I didn't ask you to leave – I asked if you *wanted to* leave."

"Yeah, but why would you say that if you didn't want me to—"

"Because I know I can be over-bearing!"

With a shrug, Naiche said, "Since when did that ever bother *me*?" She sat back down and leaned forward explaining, "I thought I was cramping your style."

"Cramping my...." Matt threw his hands up in the air, admonishing her, "You have lived with me this long and haven't figured out that I don't *have any* style?" While she was laughing, he confirmed, "So, you don't wanta move out?"

"Nope." She studied him for a second. "You don't want me to move out?" When he shook his head she said, "So, I won't move out."

"Good, that's settled."

"Careful. You may be stuck with me until I muster out in ten years or so."

That news seemed to startle him and he asked in surprise, "Why? What's gonna happen in ten years?"

"Oh, I guess I never talked about that." She took a deep breath and then explained, "I figure I'll have my kids sometime in my early forties and I wanta...I plan to raise them...Chiricahua." She studied her father closely to see if he was hurt or offended.

"That makes sense."

"Really?" she asked eagerly.

"Yeah, if they're gonna turn out anywhere near as well as you did, they'd have to be."

After they exchanged affectionate smiles she asked, "When do you think you'll muster out? Gonna go to the commercial space-flight sector someday like Bastié?"

"Nah, I'll probably end up like old Commander Weingarten – wandering The Rock, scaring first-year cadets, cutting in line at the mess hall, rambling on about how it was 'back in my day,' and being an all-around pain in the ass."

"Well, as *irresistible* as that all sounds," she joked, "alternatively, you could come with me to Chiricahua territory."

"And do what?"

"Help me build the Naomi Decker Memorial Health and Education Center."

Matt was startled speechless for a second and then said, "Oh, *that's* why you've been hoarding all of your money. You're saving up for that?"

"Yep." She grinned, offering, "And then when I'm running it, you could be raising my kids for me. After all, if they're gonna turn out like me, their grandfather has to raise them."

"What about their father?"

With a casual shrug, Decker said, "There might be a father, there might not. There's lots of ways to have kids these days."

"Tal Jacoway might have something to say about *that*," he archly objected.

"It was one date, Pop."

"One date," Ricci scoffed. "And all those nights he spent in your quarters on the way home – what do you call that?"

Naiche stared at him, open-mouthed in surprise before asking, "How do you know about that? How does *anyone* know about that? I only told Con! And I know for a fact he can keep a secret."

Ricci chuckled. "Do you *really* think those over-night visits went completely unnoticed – on *Lovelace*?" He rolled his eyes and shook his head at her, continuing, "I found out about it the same place everyone found out about me and Jackie."

"Oh. Lindstrom. Of course." After fuming for a moment, she groused, "If his appetite for food was anywhere near as big as his appetite for gossip, he wouldn't be so damn skinny." Naiche took a deep breath, offering, "Okay, there's obviously...*something* between me and Tal but it's very new and we're still figuring out exactly how deep it runs. Don't be planning a wedding just yet."

"I don't care about that – I just want you to be happy. You're happy?" When she smiled and nodded, he said, "Good," and went back to reading. Naiche had just pulled out her own hand-held to check her messages when her father said, in an off-hand manner, "Although...Zache and Nik *might* feel differently about your single status. *Especially* Nik...."

Her head snapped up. "What does that mean?" Matt just looked at her with raised eyebrows and smiled, so Naiche asked, "Oh no, is that what this dinner is about tonight? Is that why they invited Con and Aqila, too?"

"No. It's just a 'welcome home' celebration Nik is throwing." She stared at him in a silent challenge and he finally admitted, "If he happens to prod you and Tal about using the occasion to announce an engagement, well then, you simply tell him what you just told me."

Rather than answering, Deck announced, "I'm going to kill him."

"He's making that roast lamb you love," Ricci ventured, slyly.

"With rosemary potatoes?" When her father nodded, she considered the matter for a second, saying, "You know...on second thought, if Tal and I can survive the Expanse together, we can take a little teasing from Nik." Matt laughed in agreement and went back to his book. Naiche added, "After all, that's what family does."

Chiricahua Glossary

Ka dish day – until we meet again
N'daa – stranger, enemy, non-Apache people
Ndee – the people, specifically the Apache people
Shilsáásh – my friend
Shimáá – my mother
Shitaa – my father
Tzi-ditindi – sounding wood, bullroarer
Ya a teh – greetings/hello
Yadalanh – good-bye

Gą́hé, cheełkéń nts'ís dighį.Beenzhǫ́go 'ánágódlá.Gots'íse 'ádeenágódlá.Gogálei ká'ánágódlá. — Mountain Spirit, leader of the Mountain Spirits, your body is holy. By means of it, make him well again. Make his body like your own. Make him strong again.

Choctaw Glossary

Chahta imanumpa ish anumpola hinla ho? – Do you speak Choctaw?
Talako – a gray eagle

Other Exquisite Speculative Fiction from D. X. Varos, Ltd.

Therese Doucet: The Prisoner of the Castle of Enlightenment

Samuel Ebeid: The heiress of Egypt
The Queen of Egypt

G. P. Gottlieb: Battered
Smothered *(coming Summer 2020)*

Phillip Otts: A Storm Before the War
The Soul of a Stranger *(coming Fall 2020)*

Erika Rummel: The Inquisitor's Niece

J. M. Stephen: Into the Fairy Forest

Felicia Watson: Where the Allegheny Meets the Monongahela
We Have Met the Enemy

Daniel A. Willis: Immortal Betrayal
Immortal Duplicity
Immortal Revelation
Prophecy of the Awakening

CPSIA information can be obtained
at www.ICGtesting.com
Printed in the USA
LVHW031835170220
647200LV00001B/42